Technical Centre for Agricultural and Rural Co-operation (ACP-EU)

The Technical Centre for Agricultural and Rural Co-operation (CTA) was established in 1983 under the Lomé Convention between the ACP (African, Caribbean and Pacific) Group of States and the European Union Member States.

CTA's tasks are to develop and provide services that improve access to information for agricultural and rural development, and to strengthen the capacity of ACP countries to produce, acquire, exchange and utilise information in these areas. CTA's programmes are organised around four principal themes: developing information management and partnership strategies needed for policy formulation and implementation; promoting contact and exchange of experience; providing ACP partners with information on demand; and strengthening their information and communication capacities.

CTA, Postbus 380, 6700 AJ Wageningen, The Netherlands.

Agency for Cultural and Technical Co-operation (ACCT)

The Agency for Cultural and Technical Co-operation, an intergovernmental organisation set up by the Treaty of Niamey in March 1970, is an association of countries linked by their common usage of the French language, for the purposes of co-operation in the fields of education, culture, science and technology and, more generally, in all matters which contribute to the development of its Member States and to bringing peoples closer together.

The Agency's activities in the fields of scientific and technical co-operation for development are directed primarily towards the preparation, dissemination and exchange of scientific and technical information, drawing up an inventory of and exploiting natural resources, and the socioeconomic advancement of young people and rural communities.

Member countries: Belgium, Benin, Burkina Faso, Burundi, Canada, Central African Republic, Chad, Comoros, Congo, Congo (Dem. Rep.), Côte d'Ivoire, Djibouti, Dominica, France, Gabon, Guinea, Haiti, Lebanon, Luxembourg, Mali, Mauritius, Monaco, Niger, Rwanda, Senegal, Seychelles, Togo, Tunisia, Vanuatu, Vietnam.

Associated States: Cameroon, Egypt, Guinea-Bissau, Laos, Mauritania, Morocco, St Lucia.

Participating governments: New Brunswick, Quebec.

Titles in *The Tropical Agriculturalist* series

Sheep	ISBN 0–333–52310–5	Animal Health Vol. 1	0–333–61202–7
Pigs	0–333–52308–3	Animal Health Vol. 2	0–333–57360–9
Goats	0–333–52309–1	Warm-water Crustaceans	0–333–57462–1
Dairying	0–333–52313–X	Livestock Production Systems	0–333–60012–6
Poultry	0–333–52306–7	Donkeys	0–333–62750–4
Rabbits	0–333–52311–3	Camels	0–333–60083–5
Draught Animals	0–333–52307–5	Tilapia	0–333–57472–9
Ruminant Nutrition	0–333–57073–1		
Animal Breeding	0–333–57298–X		

Upland Rice	0–333–44889–8	Maize	0–333–44404–3
Tea	0–333–54450–1	Plantain Bananas	0–333–44813–8
Cotton	0–333–47280–2	Coffee Growing	0–333–54451–X
Weed Control	0–333–54449–8	Food Legumes	0–333–53850–1
Spice Plants	0–333–57460–5	Cassava	0–333–47395–7
Cocoa	0–333–57076–6	Sorghum	0–333–54452–8
The Storage of Food Grains and Seeds	0–333–44827–8	Cut Flowers	0–333–62528–5
Avocado	0–333–57468–0	Coconut	0–333–57466–4
Sugar Cane	0–333–57075–8	Market Gardening	0–333–65449–8
Oil Palm	0–333–57465–6	Forage Husbandry	0–333–66856–1
Food Crops and Drought	0–333–59831–8	Alley Farming	0–333–60080–0
		Chickpeas	0–333–63137–4

Other titles published by Macmillan with CTA (***co-published in French by Maisonneuve et Larose***)

Animal Production in the Tropics and Subtropics	ISBN 0–333–53818–8
Coffee: The Plant and the Product	0–333–57296–3
The Tropical Vegetable Garden	0–333–57077–4
Controlling Crop Pests and Diseases	0–333–57216–5
Dryland Farming in Africa	0–333–47654–9
The Yam	0–333–57456–7

Land and Life series (***co-published with Terres et Vie***)

African Gardens and Orchards	ISBN 0–333–49076–2
Vanishing Land and Water	0–333–44597–X
Ways of Water	0–333–57078–2
Agriculture in African Rural Communities	0–333–44595–3

Series Editor
René Coste
Formerly President of the IRCC

General Editor, Livestock Volumes
Anthony J. Smith
Centre for Tropical Veterinary Medicine
University of Edinburgh

Pigs

David H. Holness
Livestock consultant, Harare, Zimbabwe

with the collaboration of
Dr Noël Chabeuf
Institut d'Elevage et de Médecine
Vétérinaire des Pays Tropicaux,
Maisons-Alfort, France

CTA
MACMILLAN

The Tropical Agriculturalist Series originated under the title *Le Technicien d'Agriculture Tropicale* published by G. P. Maisonneuve et Larose, 15 rue Victor-Cousin, 75005 Paris, France, in association with the Agency for Cultural and Technical Co-operation based in Paris, France. Volumes in the series in the French language are available from Maisonneuve et Larose.

First published 1991 by
MACMILLAN EDUCATION LTD
London and Oxford
Companies and representatives throughout the world

ISBN 0–333–52308–3

12 11 10 9 8 7
08 07 06 05 04 03 02 01 00

This book is printed on paper suitable for recycling and made from fully managed and sustained forest sources.

Phototypeset by Input Typesetting Ltd, London

Printed in Hong Kong

A catalogue record for this book is available from the British Library.

Cover photograph of Landrace piglets courtesy of NHPA

Contents

Acknowledgements

This book owes much to the experience I gained over many years of visiting and discussing problems with pig producers, both large and small, throughout Zimbabwe and surrounding territories. The period I served as Director of the Pig Industry Board of Zimbabwe was particularly fruitful in this respect.

I would like to gratefully acknowledge the contributions of information, ideas and photographic material from Dr Noël Chabeuf of the Institut d'Elevage et de Médecine Vétérinaire des Pays Tropicaux in France, Dr Raul H. Godoy Montañez of the Universidad Autónoma de Yucatán in Mexico, Dr Nimal Pathiraja of the University of Zimbabwe, Mr Rex Parry of Macmillan Education, Dr Tony Smith of the Centre for Tropical Veterinary Medicine in Edinburgh, Scotland, and the staff of the International Livestock Centre for Africa in Ethiopia. Other technical contributions have been acknowledged in the text.

I am particularly indebted to Ms Lyn Purves who read the manuscript and made many helpful suggestions and corrections. My grateful thanks also go to Ms Grace Kahari and Ann Smythe for their typing and help with the preparation of the manuscript, and to Ms Janine Holness for the original illustrations.

The author and publishers would like to thank Dr S. H. B. Lebbie of the International Livestock Centre for Africa (ILCA), Addis Ababa for advice and comments on the manuscript, and Dr R. T. Wilson, also of ILCA, for arranging Dr Lebbie's review. Thanks are also due to the author and publishers of Whittemore, *Elements of Pig Science* (Longman, 1987) for permission to redraw their Figure 2.16, p. 30.

Photographs courtesy of R. Matthewman (Fig. 17) and A. J. Smith (Fig. 20). All other photographs provided by the author.

Preface

This book is the third in a series of fifteen books (the other two were on poultry and sheep) to be produced by CTA/Macmillan in the field of animal production in the tropics. The series has been written to give up-to-date information to both students and farmers. All the books have been written by specialists who have worked in a number of tropical countries or regions.

Over the past forty or more years pig production in Western countries has become more and more intensive. Emphasis has been on output per man employed and output per unit of capital invested. Few books have been addressed to the problems faced in producing pigs under tropical and subtropical conditions. Under these latter conditions the problems facing the pig-keeper are very different to those facing his counterpart in developed countries. Apart from the obvious climatic differences there exist chronic problems in obtaining adequate feed supplies; at the same time inadequate control of disease may make intensive pig production unprofitable or even untenable. Religious considerations may make pig-keeping unacceptable and the possibility of transfer of disease and parasites to the human population may make extensive pig production unwise.

This book attempts to address these problems and to see which systems of pig production are most acceptable in various regions of the tropics and which should be avoided. It approaches the subject from the point of view of both the commercial producer and of the village pig-keeper. It aims to be readable, informative and to transfer ideas not only from developed countries to the tropics but also between tropical countries: this latter form of technological transfer being one of the most useful but also one of the most ignored.

Anthony J. Smith

1 Distribution, potentials and constraints

World pig population

The estimated world pig population of 826 million (FAO, 1988) means that there is approximately one pig for every six people in the world. Although pigs are numerically fewer than some other domestic species, more pigmeat is produced than any other meat (Table 1.1). This reflects the greater productivity of the pig when compared with other domestic species.

Table 1.1 A comparison of the main livestock species in the world in terms of numbers and meat production

	Numbers (million head)	Meat output (000 metric tonnes per year)
Cattle	1 253	50 098 (Cattle and Buffalo)
Buffalo	137	
Sheep	1 174	8 801 (Sheep and Goats)
Goats	521	
Poultry	10 050	11 495
Pigs	826	63 917

(Source: FAO *Quarterly Bulletin of Statistics, 1989*)

Distribution and consumption

The distribution of pigs throughout the world is not uniform. Nearly half the world's pig population is in Asia, with a further 30 per cent in Europe and the USSR. In contrast, the population in large parts of the tropical and sub-tropical developing regions (e.g. Africa and Latin America) is relatively small (Figure 1). Nevertheless, the increase in the world pig ver the last decade is largely attributable to increases within ing world, which now constitutes some 60 per cent of the ulation of pigs (Figure 2). It is noteworthy that the majority gs in the developing world are located in one Asian country, China.

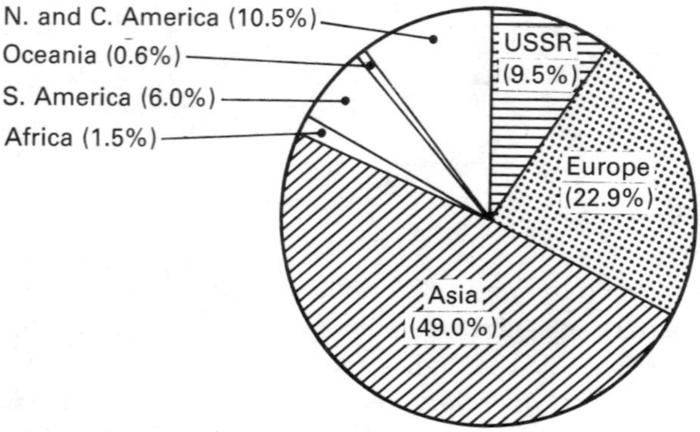

Fig 1 *Regional distribution of the world pig population (FAO 1988)*

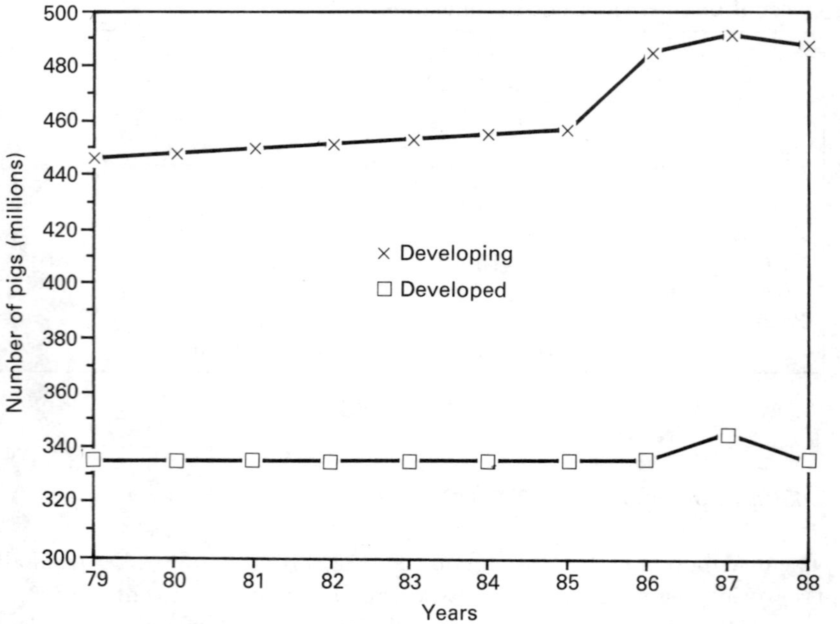

Fig 2 *Increases in the pig population in developed and developing regions of the world (1979–88)*

Similarly, marked differences exist in the consumption patterns of pigmeat throughout the world. In some parts of Europe, annual per capita consumption of pigmeat is over 50 kg, and represents some 60 per cent of the total meat consumed. At the other end of the scale, in areas of the developing world, and particularly in Africa, estimated annual per capita consumption ranges from 1 to 3 kg, and forms less than 10 per cent of the total meat diet.

The reasons for the uneven distribution of pigs throughout the tropical and sub-tropical world are manifold. In tropical Asia and parts of China, pork is the predominant component of the diet. On the other hand, in areas where the Islamic religion prevails, e.g. the Middle East, Pakistan and parts of Africa, Muslims are forbidden to eat any pig meat. Similarly, believers in the Jewish faith are instructed not to eat pork meat, and Zionist sects occur throughout the developing world. Social factors also play a part and these may have a positive or a negative effect on the pig population. In some Pacific islands, such as Tonga and Papua New Guinea, pigs are highly regarded as a source of wealth and associated with marriage customs. On the other hand, in Africa people have traditionally obtained their meat supplies mainly from ruminants, particularly cattle, and this preference persists.

The pig has historically been considered an unclean animal, wallowing in filth, an object of distaste and a hazard to human health. Clearly there is some truth in this assumption if the pig is used as a scavenger, but the exact opposite pertains if the pig is well managed under confined conditions.

Climate has an influence on distribution. Pigs can be reared almost anywhere given suitable housing and management, but in situations of extreme temperatures, humidity or lack of rainfall they cost more to produce, because of the need for more expensive housing and because suitable foods may not be available.

The potentials and constraints to pig production in developing countries

The world trend is towards the consumption of more white than red meat. Thus the potential for increased meat production from pigs in the developing world is enormous. When compared with cattle and other ruminants, pigs have some major potential advantages, namely:

- they produce meat without contributing to the deterioration of the natural grazing lands. This is of paramount importance in relation to the current steady desertification, soil erosion and loss of productive land in tropical and sub-tropical parts of the world. Overstocking and consequent overgrazing by ruminants is a primary cause of this degradation.

- they convert concentrated food to meat twice as efficiently as ruminants.
- they possess the potential to be highly productive. Because they are capable of producing large litters after a relatively short gestation period, and have a short generation interval and grow rapidly, their output in terms of yield of meat per tonne of liveweight of breeding females per year is in the region of six times that of cattle.
- if confined, maximum use can be made of their manure and effluent (see Chapter 3).
- their relatively small size, when compared with cattle, provides for more flexibility in marketing and consumption.
- the meat pigs produce is particularly suitable for processing; some of the processed products have a longer shelf life than fresh meat, and can thus be distributed to a wider section of the population.
- quicker turn-over rate on investment compared with cattle.

Apart from the social and religious constraints mentioned, other constraints to pig production are:

- as simple-stomached animals, they compete directly with humans for food, especially the staple grains and oilseeds. This can be partly overcome by making maximum use of crop by-products, waste feeds and grain unsuitable for human consumption.
- they cannot provide a source of draught power for farming operations.
- since they tend to be raised close to human habitation, their effluent may cause a pollution problem.
- because pigs and man are co-hosts to a number of parasites, if pigs are not confined they can pose a problem to human health.

2 Understanding the pig

Introduction

It is essential for anybody involved in the breeding and rearing of pigs to have a basic understanding of some important aspects of the biology of the animal. This leads to an appreciation of the likely husbandry requirements of the pig.

Origins and evolution

Although the actual origins of the domestic pig are obscure, it is probable that it is mainly derived from the European wild boar (*Sus scrofa*). Originally pigs colonised the forests and swamps (as does *Sus scrofa* to this day), and were thus designed to live in a moist, shady environment. Their short legs and powerful streamlined body were built for moving through dense undergrowth, and the strong head and tusks, with a cartilaginous disc in the snout, for digging and rooting.

Biblical writings indicate that pigs were first domesticated as early as 2000 BC. As man has developed the pig as a meat animal, major changes in conformation have occurred from the typical, 'unimproved' type. The relatively large, narrow head, heavy forequarters, tapering light hindquarters and compact body have been replaced by a smaller head, lighter forequarters, a longer and wider body with bigger capacity and well-developed, meaty hindquarters (see Figure 3).

Feeding and digestion

The first 72 hours after birth are very critical for the baby pig. During this period the colostrum of the sow has a high content of antibodies and the piglet intestine is able to absorb intact proteins. As the piglet has very little of its own resistance to disease, it is essential that it gets a good suck of colostrum and acquires passive immunity from the sow.

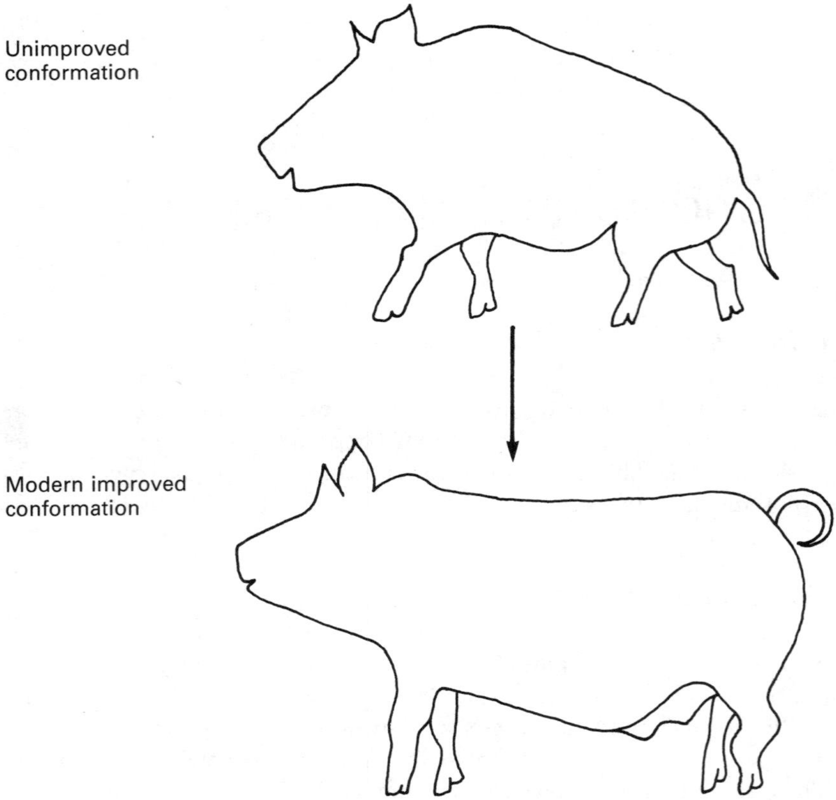

Fig 3 *The change in body conformation as a result of selection for meat production*

Failure to take in sufficient colostrum will invariably result in the pig succumbing to infection before it can develop active immunity of its own. Once the piglet has established a teat position, which normally occurs in the first 24 hours after farrowing, it will retain this position for the remainder of the suckling period. As long as milk production is adequate, a sow will suckle her litter every 60 to 90 minutes.

Alimentary canal

Although pigs in tropical regions may eat a lot of fibre they are simple-stomached animals and not ruminants which possess a complex stomach with a large microbial population which enables them to digest large quantities of fibrous material (Figure 4). Thus, their ability to digest and utilise fibre is restricted to that digested by the microbial population in the caecum which is of relatively small volume when compared with the rumen (see Figure 4). It has been claimed that unimproved breeds found in Africa have an enhanced ability to utilise fibrous foods compared with exotic breeds. While this may be so to a small extent, there are no anatomical differences of the digestive tract between the two types. Accordingly, with all pigs, high-fibre diets will have the effect of diluting the amount of nutrients available to the animal. In contrast to ruminants, pigs are unable to utilise non-protein sources of nitrogen for the production of microbial protein in the rumen. This makes them dependent on both the amount and quality of protein in their diet.

The alimentary tract of the pig (Figure 4) is designed to digest and absorb concentrated foods. Food taken in at the mouth is ground into a pulp by mastication. At the same time it is moistened and mixed with saliva. Saliva contains the enzyme ptyalin which initiates the breakdown of starch to simpler carbohydrates. The food then passes on into the stomach, which provides an acid environment due to the presence of hydrochloric acid. The gastric juice contains the enzyme pepsin which begins the breakdown of protein.

The small intestine is the major site where food absorption occurs, and

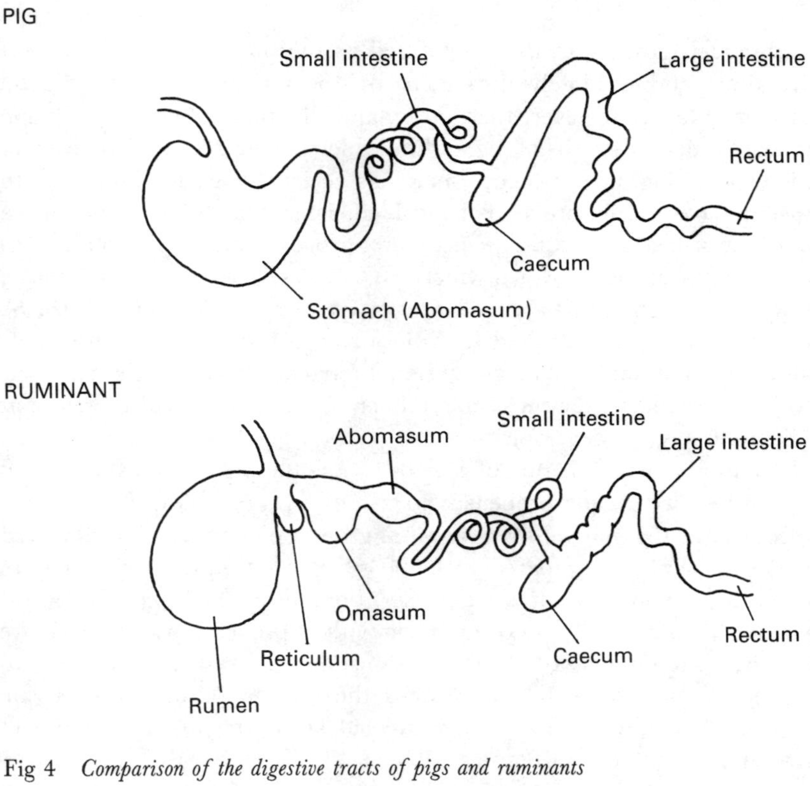

Fig 4 *Comparison of the digestive tracts of pigs and ruminants*

digestive juices from the pancreas, liver and the small intestine complete the process of digestion as follows.

Starch is hydrolysed to maltose by amylase from the pancreatic juice. Maltose and other disaccharide sugars are broken down by specific enzymes in the intestinal juice, e.g. maltase, lactase and sucrase into monosaccharides such as glucose and fructose. These are then absorbed through the gut wall.

Trypsin in the pancreatic juice acts on protein to produce polypeptides, which are then broken down to amino-acids by various peptidases in the intestinal juice and subsequently absorbed.

Bile, which is secreted by the liver, serves to emulsify fats into smaller globules, which are then broken down by the enzyme lipase into fatty acids and glycerol ready for absorption. Lipase is present in both the pancreatic and intestinal juices.

Pigs are omnivores and will consume a wide range of foods from both plant and animal sources. The natural inclination of the pig is to eat on a 'little and often' basis, and this is likely to maximise both total food intake and the efficiency of food utilisation.

Growth and development

In practical terms, growth is measured as the increase in body weight with time, and is largely dependent on the amount of food, or total nutrient, intake. However, there are major differences between the food intake of different breeds of pig and this affects their growth response per unit of food ingested. Because man has selected pigs for high growth rates in order to improve biological efficiency, he has selected for a larger mature size. In consequence, unimproved types of pig common in developing parts of the world which have not been selected for increased growth rates will tend to grow more slowly to a smaller mature size (A in Figure 5) when compared with improved breeds (B in Figure 5). It follows that if unimproved pigs are slaughtered at the same weight as their exotic counterparts, they will be relatively more mature and therefore at a different stage of development.

Equally important as rate of growth is how the pig develops. Selection has resulted in a greater propensity to lay down protein tissue in improved breeds. Thus the plateau for maximum growth potential in an improved breed can be 600 g compared with 400 g for an unimproved pig (Figure 6). As the level of feed intake increases, the unimproved pig will deposit more fat (A in Figure 6), in comparison with improved types (B in Figure 6). Because too much fat is neither a consumer desirable, nor is it cheap to produce (approximately five times the nutrient cost of lean tissue deposition), it is critical that pigs are fed according to their ability to grow and lay down lean tissue.

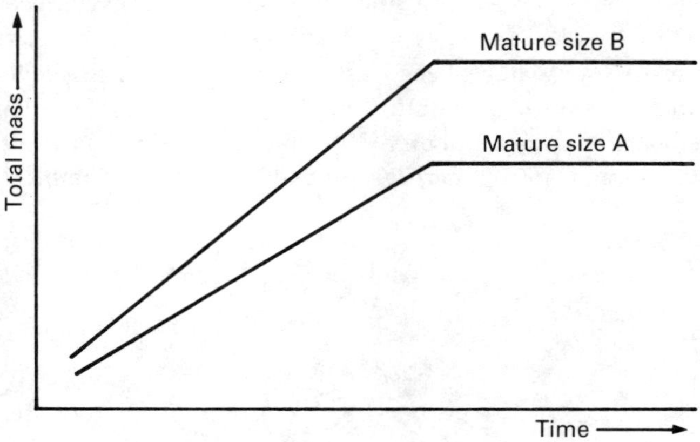

Fig 5 *Differences in mature size and consequences for growth rate in unimproved (A) and improved (B) breeds of pig*

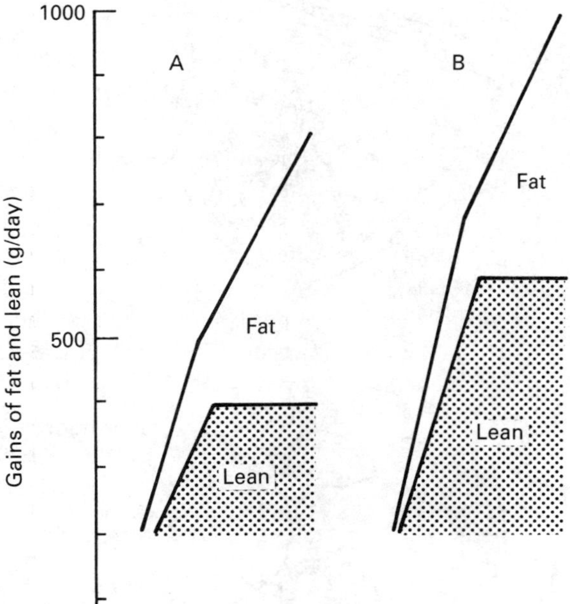

Fig 6 *Typical differences between unimproved (A) and improved (B) pig breeds in their ability to lay down lean and fat tissue as food intake is increased (After Whittemore, 1987)*

Entire male pigs grow faster, have leaner carcasses and convert feed more efficiently than females. If males are castrated the case is exactly opposite. Traditionally, pigs have been castrated in order to improve carcass quality and to prevent boar 'taint' or odour in the meat which tends to occur as boars approach puberty. Nowadays, modern pigs grow faster and are slaughtered at younger ages and the problem of taint is

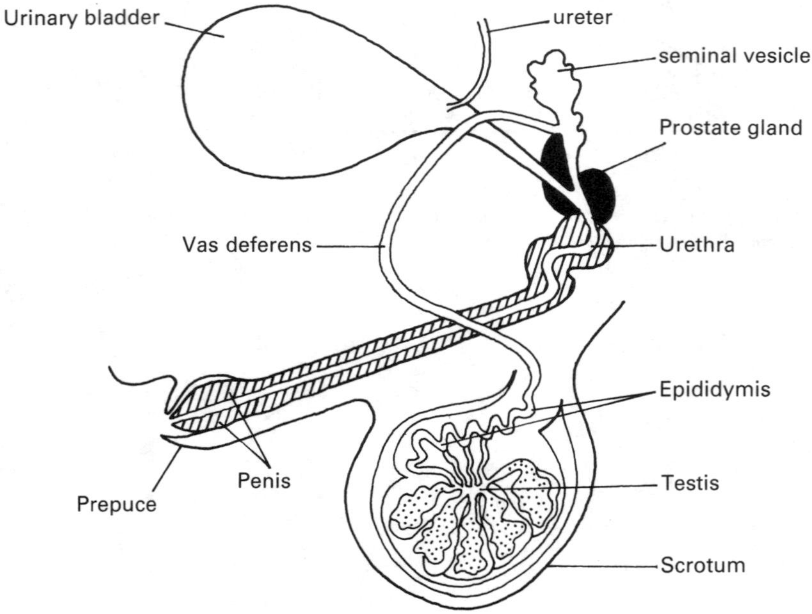

Fig 7 *The male reproductive system*

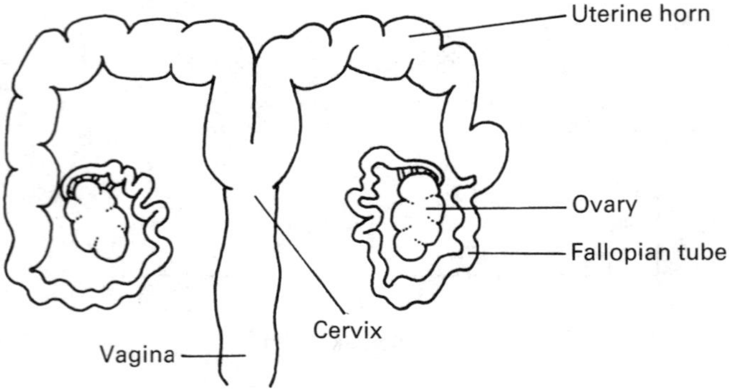

Fig 8 *The female reproductive tract*

considerably reduced. Unless pigs are grown slowly or are required for a highly sophisticated market, there would appear to be no justification for castration in pigs destined for meat production.

Baby pigs are born with less than 2 per cent of fat in their bodies, which makes them particularly susceptible to cold stress. Thereafter they deposit fat rapidly, and will usually have a body fat level of over 15 per cent by the time they are three weeks old. This serves as a reserve of energy as they adapt to a reduction in milk intake and to overcome the stress associated with weaning.

Reproduction

Males

The male reproductive system (see Figure 7) is characterised by a pair of relatively large testes, which can weigh over 300 g each in some exotic breeds. Together with the secretions from the accessory sex glands, the testes can produce up to a litre of semen in a single ejaculate.

To facilitate the transfer of these large quantities of semen at coitus, the end of the penis of the boar is spiral in shape which enables it to lock into the cervix of the sow. The duration of coitus varies but may last for 20 minutes.

Puberty, or the ability of the boar to serve a sow, generally occurs around four months of age, but may be earlier in unimproved breeds. However, boars should not normally be used until seven months old. Young boars are susceptible to bullying by mature sows, and this may adversely affect their subsequent mating performance.

Females

The female reproductive tract is distinguished from other farm species by the long, convoluted uterine horns (700 to 800 mm in length), which are designed to accommodate large numbers of foetuses (Figure 8). The sow will ovulate simultaneously from both ovaries, normally shedding between 11 and 24 eggs.

Puberty, marked by the onset of oestrous cycles, occurs between five and seven months, but may be as early as three months in unimproved breeds. The number of eggs shed at ovulation, and therefore potential litter-size, increases gradually over the first few oestrous cycles.

The sow will cycle and show heat every 21 days (range 18–24). She will not cycle when she is either pregnant or lactating, although sows will sometimes show heat during lactation when run in groups. Heat lasts from one to three days, and ovulation occurs by the second day of oestrus or any time thereafter.

After coitus and fertilisation, the embryos space themselves evenly throughout the entire uterus before implantation. Competition for space, nutrients and other unknown factors often results in uneven growth rates *in utero*, which gives rise to variation in piglet birthweight. The lighter pigs then suffer a disadvantage in the competition of early post-natal life. This problem tends to be accentuated in older sows, due to the effects of wear and tear on the uterus.

Pregnancy lasts for 114 days but will tend to be extended slightly with smaller litters. Farrowing may vary in duration from two to 24 hours and will tend to be longer the more piglets that are produced. However, due to the relative difference in size between piglets and the dam, and the type of placentation in sows, farrowing is normally a straightforward process. The incidence of stillborn piglets, which may be due either to death *in utero* or during the birth process, is greater in larger than in smaller litters.

Overall, reproduction in the sow results from a complex hormonal interplay between the brain, the pituitary gland, the ovaries and the uterus. These complex relationships must be borne in mind by the pig-keeper when management strategies are designed, otherwise optimum reproductive performance by the sow will not be achieved.

Behaviour

Pigs are not solitary animals and will generally benefit from social contact with each other, even if only by sight or smell. At the same time, groups of pigs will always establish a social hierarchy and this starts at birth when the piglet struggles to commandeer a teat position. If strange pigs are penned together later in life, fighting invariably ensues, and this can lead to considerable stress and physical damage to individual pigs. Once settled, however, pigs will huddle together in order to retain body heat in cold weather.

In common with the female of other species, sows are notably more docile during pregnancy than when they have just produced young. Just prior to farrowing, the sow will prepare a nest from her bedding. She is often irritable during this period, and if she is confined without access to bedding material, this can lead to stress during the farrowing process.

Contrary to popular belief the pig is not a dirty animal and will normally defaecate and urinate away from its resting or lying areas. However, this pattern tends to break down if pigs are overcrowded or stressed in other ways. Also, when temperatures are high they will often roll in their own faeces and urine in an attempt to increase evaporation and keep cool.

Recent studies have highlighted the importance of the interaction between pigs and humans in relation to productivity. If pigs live in fear of their stockman, both growth and reproductive performance are likely to be depressed.

The effect of climate and temperature

In order to cope with forest conditions, the pig has better mechanisms for retaining heat, especially a well-developed subcutaneous fat cover, than for losing heat from the body. Because the pig possesses sweat glands only on the snout, it is unable to dissipate large amounts of heat by sweating. Furthermore, the skin of certain breeds of pig, e.g. Large White and Landrace, has no protection against the sun, and unless they have access to shade, or mud in which they can wallow, they can become badly sunburnt.

Like man, the pig is a homeotherm, and needs to maintain its deep body temperature constant. Nature has designed the metabolism of the pig to operate most effectively at 39°C. For a certain range of environmental temperature, known as the zone of thermal neutrality (see diagram in Figure 9), the pig finds this easy and can maintain the correct body temperature by varying blood flow to and from the skin. The extent of this zone changes quite markedly according to the weight of the pig (Figure 10).

At the bottom end of this zone (point Y in Figure 9) a lower critical temperature is reached when the pig requires to divert food energy to

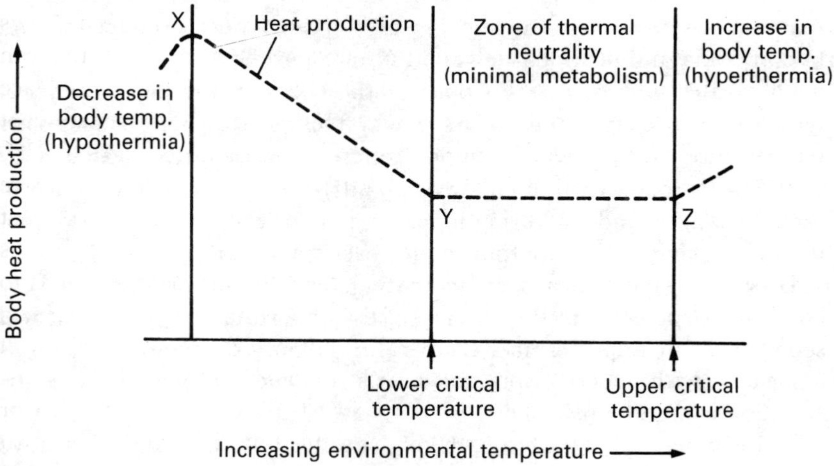

Fig 9 *The influence of environmental temperature on body temperature and heat production in the pig*

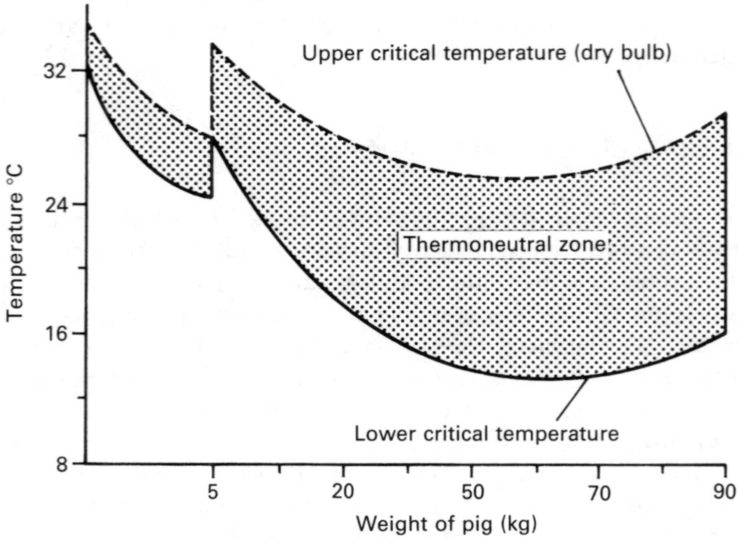

Fig 10 *The change in the extent of the thermoneutral zone with weight of the pig*

increase heat production in order to maintain body temperature. The lower critical temperature will vary between pigs according to a number of factors, for instance how fat (well-insulated) or thin the pig is, how much food it is eating and therefore how fast it is growing, whether it has bedding to help prevent heat loss, whether it can huddle with pen-mates, and whether it can make postural changes to minimise heat losses. Eventually, with decreasing ambient temperature the pig can no longer maintain its body temperature in spite of high heat production, and hypothermia and death ensue (Point X in Figure 9).

Of greater interest in the tropics is the effect on the pig as ambient temperature rises. When the environmental temperature approaches body temperature, the pig will attempt to increase evaporative heat loss by sweating (through its limited sweat glands), panting, postural and positional changes, and wallowing in water, mud or excrement. In addition, it will reduce its energy output by decreasing its feed intake.

However, as the means of dissipating heat in the pig are not very efficient, particularly in the absence of the opportunity to wallow, it will soon reach an upper critical temperature (Point Z, Figure 9). This is associated with hyperthermia and heat stress and the pig will die if the situation cannot be reversed.

At the other end of the scale, the piglet at birth is particularly sensitive to low ambient temperatures. Pigs are born with virtually no subcutaneous fat cover and limited carbohydrate reserves and therefore at birth they will suffer an immediate drop in body temperature. In the case of weaker piglets they may battle to obtain an adequate milk supply; if they require energy to keep warm they very quickly develop the condition of low blood sugar (hypoglycaemia), and die of cold.

Stress

From the foregoing it is clear that stress factors can take many forms, and in fact can involve fear, pain, temperature, direct sunlight, restraint, fatigue and interference with natural behaviour patterns. Stress will quickly lead to reduced performance and productivity, and specifically to gastric ulcers (just as in humans), greater susceptibility to infectious diseases and higher mortality rates. It is therefore paramount that we understand what constitutes the major stress factors in pigs in different circumstances so that production systems can be designed to minimise these effects.

3 Systems of production

Small-scale systems

Pigs kept as scavengers

This is the traditional system of rearing pigs in large parts of the tropics (Figure 11, Stage 1). It is also the simplest and cheapest. Each family, kraal or village keeps a few pigs which are allowed to wander freely and pick up food when and where they can. If extra food is available, they may receive supplementary food. This supplementary food will generally consist of foods of low nutritional quality such as banana and maize stalks, water hyacinth, rice bran, local herbaceous plants, by-products of beer-making or kitchen wastes. Where pigs are particularly valued for ritual slaughter, as in parts of Asia, a few may be confined and fed for a three-to-six month fattening period prior to important ceremonies.

Indigenous breeds of pig predominate on the scavenging system, because they are adapted to the local environment and their relatively small size and mobility render them best able to cope with the conditions (see Figure 12). Productivity is normally low with the erratic and often seasonal food supplies resulting in irregular breeding of sows, high rates of piglet mortality and low growth rates. Pigs raised on this system are particularly susceptible to infestation with parasites, and invariably carry a heavy burden of intestinal roundworms and tapeworms. A particular hazard of the scavenging system is that pigs have access to sources of parasites, e.g. human excreta, which can then be transmitted back to man when he eats the meat (see Chapter 7).

The majority of scavenger pigs are owned by subsistence farmers, and are not produced with any particular market in mind. Rather the pig plays an important socio-economic role as a form of bank, and pigs are sold at times of cash shortages or unexpected needs in the family. The numbers are usually kept small, in the region of 1–3 breeding females per herd.

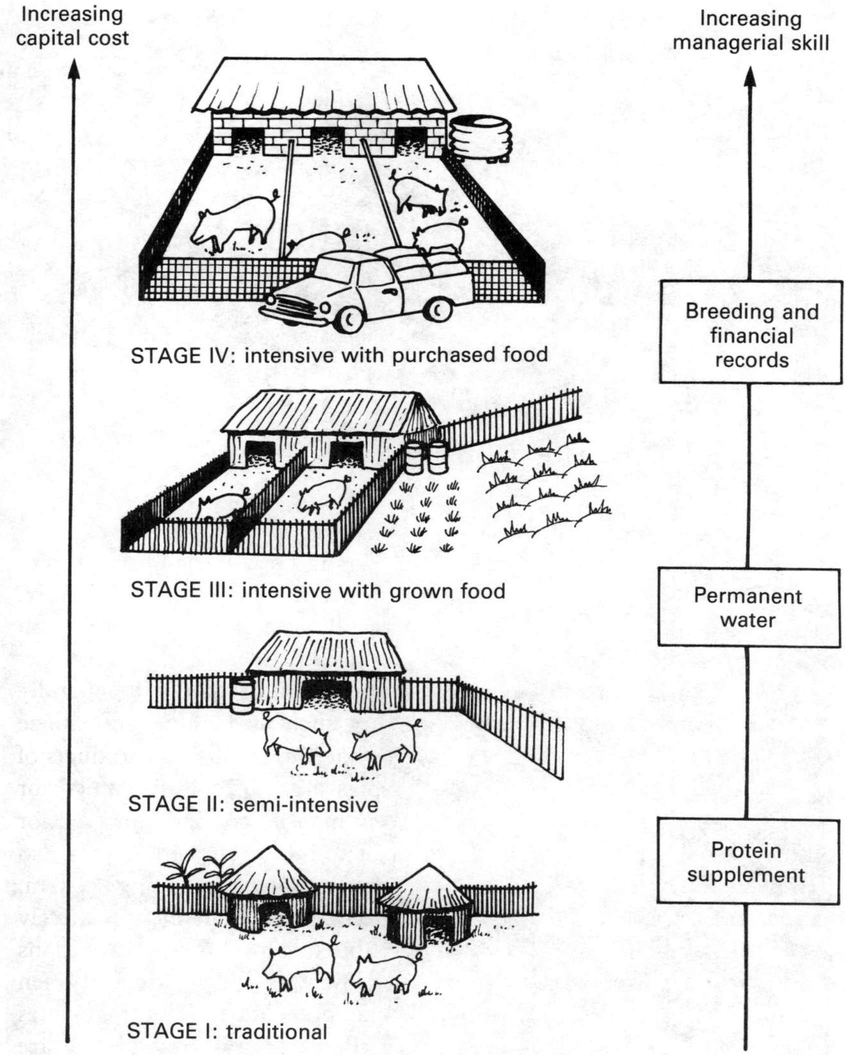

Fig 11 *The ladder of piggery development*

Semi-intensive production

In these systems, often also known as 'backyard' systems, pigs are confined and therefore represent a commitment on the part of the farmer to feed his pigs (Figure 11, Stage 2). Pens or sties are mainly of very simple construction (Figure 13), and in some countries of South East Asia and West Africa, may be made of bamboo and elevated. On the other hand, pigs are sometimes constrained by tethering in larger yards or paddocks.

Fig 12 *Indigenous pigs run on the scavenger system of production*

Fig 13 *Simply-constructed pig pens used on semi-intensive systems*

Feeding is based on kitchen waste, vegetables and by-product foods, and management is generally minimal. As a consequence, productivity tends to be relatively low and mortality can be high. Although local and indigenous pigs predominate, crosses between indigenous and exotic breeds can be found in this system of production throughout the develop-

Fig 14 *A small-scale intensive pig-production unit*

ing world. Marketing is largely indiscriminate and is dictated by the immediate financial needs of the owner. Generally, herd sizes and productivity tend to be higher in these systems than in the scavenging systems.

Intensive production

These systems are characterised by the fact that the small-scale producer has moved away from subsistence production to the commercial production of pigs. Units may comprise up to 50 head, and the producer will grow and/or purchase food specifically for his pig enterprise (Figure 11, Stages 3 and 4). The system of housing will be more sophisticated and will generally comprise a concrete or hard floor, adequate shelter, shade and pen space, and appropriate feed and watering facilities (Figure 14). In order to justify the increased capital cost, the farmer will attempt to manage his pigs to optimise output, including some veterinary protection against parasites and disease. The type of pig raised will tend to be mainly the higher-performance exotic, or a cross between exotic and indigenous breeds. Marketing may be informal, through local butchers, or into the large-scale commercial sector, but in any event will be planned to bring in a regular income for the enterprise.

Large-scale systems

Intensive

This is the most common system of large-scale production. Units are generally capital intensive and may involve sow herds from 40 up to 1000 head. Modern high-performance breeds of pig, or hybrids, are used, and provided as far as possible with optimum conditions of housing, feeding and management in order to ensure maximum output. Housing will often be designed specifically for the different classes of stock and environmental conditions.

These units, especially the larger ones, are particularly amenable to integration with grain production and stockfeed manufacturing operations on the one hand, and processing and marketing on the other. The pigs will invariably be marketed through a processor in order to maximise returns on the carcass.

Extensive systems

There is a trend throughout Europe and America towards less intensive systems of production, particularly for sows. These systems, often known as 'outdoor' systems, entail keeping sows in paddocks and providing

Fig 15 *Wallows are essential for extensive systems of production in hot conditions*

individual huts or arks for farrowing and shelter. Weaner pigs are generally raised under more intensive conditions. In Europe, hybrid sows, with a greater ability to withstand climatic vagaries, are produced specifically for use on this type of system.

Production systems of this type exist in the tropics, and there is clearly potential for further expansion. The major advantages when compared with intensive systems are that less capital is required for establishment, and sows can gain access to bulky foods such as pastures, crop residues, cassava roots and sweet potatoes. In tropical regions it is essential that adequate shade and wallows are provided (Figure 15). Moreover, there must be tight control of parasites and adequate fencing to prevent contact with endemic diseases, e.g. in Africa, contact with the Bushpig and transfer of African Swine Fever.

Integrated systems

The integration of pig production with other ancillary enterprises has long been widely practised in tropical Asia, involving in various combinations, fish farming, production of algae, methane gas generation, ducks, water hyacinth and vegetable production. Such bi- or tri-commodity operations enhance the efficiency of resource use and increase output for the overall operation.

By fertilising fish ponds with pig manure and effluent, algae are generated which can then be utilised by fish. As long as sufficient water is available for suitable fish ponds, pig sties can either be constructed above the ponds so that the manure can drop straight into the water, or close by so that the effluent can be channelled into the ponds. The various species of *Tilapia spp* are the most commonly-used fish, often mixed with small populations of carp (*Cyprinus spp*) and catfish (*Clarias spp*) or other predators. Fifty to 60 pigs produce sufficient effluent for one hectare of fishpond, which, if stocked at between 20 000 and 50 000 fish per hectare, can produce annual edible fish yields or 3·5 to 5 tonnes per hectare.

In some countries, water hyacinth is used to harvest nutrients from the fish ponds, and this is then fed back to the pigs. Alternatively, the nutrient-rich water can be used for irrigating vegetables, or in other systems, ponds can be dried in rotation and vegetables grown in the dry pond beds. In another alternative system, where other food is available, the pig effluent is used to produce algae which is then harvested and dried and fed back to pigs or other livestock. If the pig effluent is insufficient for any particular system, ducks can be used to augment the fertilisation of the ponds with their manure.

The solid fraction of pig manure can be used as an effective fertiliser for crops, particularly if it is properly composted, and this may prove to

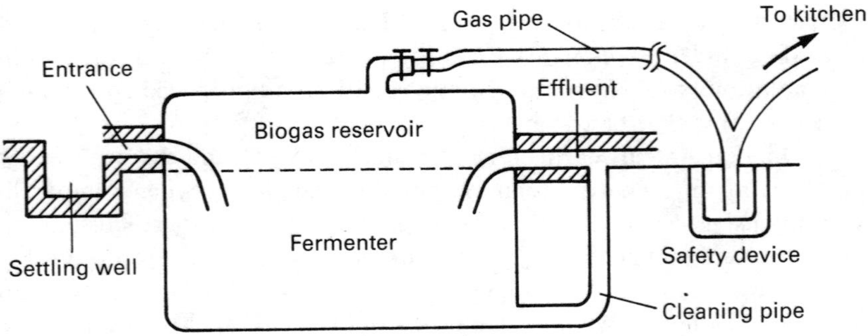

Fig 16 *A simple digester for methane production*

be the most cost-effective use of the by-product. The liquid effluent, however, is difficult to handle for fertilising arable crops, and is generally wasted if not used for other purposes. A further important development has been the anaerobic fermentation of pig effluent for the production of methane gas. Relatively simple digesters (see Figure 16) can produce a steady source of methane which can be used as a means of energy for domestic or agricultural use. These digesters are now in use in rural areas throughout the developing world, and it is estimated that seven head of pigs will provide enough dung to ensure the supply of sufficient methane for house fuel for a family of five.

4 Breeds and breed improvement

Breed types

Although some are numerically small in number, there are over 90 recognised breeds and an estimated 230 varieties of pigs in the world. They can be broadly classified into indigenous or unimproved types (see Chapter 2) or the more modern exotic types which have been selected and developed for specific commercial purposes.

Indigenous breeds (Domesticated)

These predominate in the tropical and developing world and have evolved a variety of shapes and sizes in order to survive in a range of different environments. In general, they are smaller and shorter of leg than exotic types (mature weight of females 40–120 kg), with the typical unimproved conformation of a large head, well-developed forequarters and relatively light hindquarters. This renders them more mobile and better able to forage and root for themselves. They are early sexually maturing and females may show first oestrus as early as three months of age. There are many variations of coat colour, but black and brown are most common and white is infrequent. The degree of hairiness also varies, and both hairless and relatively long-haired types are found. Within the major regions of the tropics, the main breeds and types are as follows.

Africa

In many countries in Africa, pigs have not been characterised into specific breeds, and are variously referred to as 'indigenous', 'local' or 'unimproved' pigs. The situation is further confused in some areas by interbreeding with imported exotic strains.

In North Africa, pigs are not common, and where they do occur as in Egypt, Tunisia, Morocco and the Sudan, they are of extremely primitive type with long thin heads, small prick ears and slender poorly-muscled bodies. Similar primitive types have been reported in parts of West Africa, such as Zaire and Angola. However, in some parts of Africa, more specific use is made of pigs for meat production and the Bakosi in the

Cameroons and the Ashanti Dwarf in Ghana are typical and have been described as indigenous breeds. They are small with a mature female size of 40–60 kg, and are mainly black in colour with prick ears (Figure 17).

Although the unimproved pigs of East, Central and Southern Africa are almost certainly descended from stock introduced by early European travellers, and therefore not truly indigenous, they are relatively widespread. In Tanzania, some 60 per cent of the pig population is classified as indigenous, and some quite different types have evolved in various

Fig 17 *Scavenging pigs in Ghana*

Fig 18 *Indigenous pigs found in Southern Africa*

rural areas. The indigenous pig which has been described in South Africa has been classified by Mason and Maule (1960) into two distinct types. Firstly, there is the Windsnyer type, which is a long-nosed, razor-backed pig which also occurs throughout Zimbabwe and in parts of Mozambique and Zambia (Figure 18), and secondly the Kolbroek, which is a short fat animal with a short snout and dished face, which appears to be restricted to South Africa.

Clearly the productivity of these unimproved breeds in Africa is influenced by their environment, and some typical reproductive performance figures are shown in Table 4.1. In trials in Zimbabwe, although litter sizes tended to be smaller, total liveweight of the litter as a proportion of the weight of the sow at farrowing was of the same order as exotic sows (11 per cent). The efficiency of conversion of food to bodyweight in indigenous sows is also equivalent to that of exotic types. Indigenous sows showed excellent mothering ability, which resulted in very low piglet mortality without sophisticated housing. However, during the growth phase, growth rates and food conversion efficiencies of indigenous pigs are below their exotic counterparts (Table 4.2).

Table 4.1 Some fertility and performance figures for indigenous sows in Africa

	Nigeria	Zimbabwe	South Africa	Ghana (Ashanti) (Dwarf)
Litter-size at birth	6·5	7·9	7·2	6·3
Litter-size at weaning	5·5	7·5	–	–
Pre-weaning mortality (%)	15·0	5·0	–	–
Average weaning age (weeks)	9·0	8·0	8·0	8·0
Average weaning mass (kg)	–	7·6	9·0	7·0

Table 4.2 A comparison of growth rate and food conversion efficiency of Zimbabwe indigenous and Large White pigs fed a commercial diet from 8 to 32 weeks of age

	Period (weeks) 8–16	17–24	25–32
Bodyweight gain (kg/day)			
Indigenous	0·28	0·51	0·45
Large White	0·44	0·58	0·68
FCE (kg food per kg bodyweight gain)			
Indigenous	3·1	3·6	5·0
Large White	2·7	3·3	4·0

Fig 19 *One of the many breeds of pig indigenous to China, typical of the Chinese type*

South East Asia and Southern China

The region is dominated by China, where the pig has been the major domesticated meat animal for centuries. It is estimated that there are over 100 different pig breeds in China mainly derived from the *Sus vittatus* type and these have been described in detail by Epstein (1969). The role of the pig in Chinese agriculture, with every peasant farmer keeping one or two sows, has been to dispose of farm by-products and household waste. This common system has led to many similarities between the different breeds, including docility, an ability to digest and utilise low-quality diets, and a tendency to carry a lot of fat. Generally, Chinese breeds are typified by a short dished head, a broad short body often with a hollow back and pendulous belly, and short legs (Figure 19).

The South China breeds tend to be smaller than the breeds found in Central China, and are less prolific, but piglet mortality is very low. Throughout South East Asia, pigs are derived from the Chinese breeds and the sway-back type is common. In Thailand, for example, the local genotypes are believed to be derived from the Hainan breed of mainland China (Epstein, 1969). Research in the highland areas has shown typical litter sizes of 7·1 piglets born and 5·8 weaned per sow. The indigenous pigs of Malaysia and Singapore are very similar to the South China type, and other variants are found in Sarawak, the Philippines and Sri Lanka.

South and Central America

Indigenous pigs are considerably more numerous than they are in Africa, and consequently more specific breed types have developed and been described. Most of these pigs are of the short lard type, and probably originated from Chinese pigs introduced in the fifteenth or sixteenth century.

In Central America, the small black hairless Pelon breed is common, and occurs in Mexico, El Salvador and Costa Rica. Mature females weigh about 70 kg and their productivity is relatively low. Other variants of this type occur throughout the region. An example is the Yucatan pig, native to the Yucatan Peninsula in Mexico, which is smaller, slate-grey in colour and tends to have a shorter snout. Further south in Colombia, parts of Venezuela and Bolivia, the black Criollo is a distinct breed, with longish hair and a long, narrow snout. However, in Bolivia, pure Criollo pigs are now very rare due to indiscriminate breeding with imported exotic types and subsequent interbreeding. This has resulted in a whole range of colour patterns in the village pigs. These Criollo pigs are also found in the Caribbean islands.

Pigs play a relatively important part in the economy of Brazil, and there are several indigenous breeds. These include the Pirapitinga, which is a small pig with very little hair, the Piau, which is black with white spots, and the Pereira and the Canasta, which are lard types, black or grey in colour, with considerably more hair.

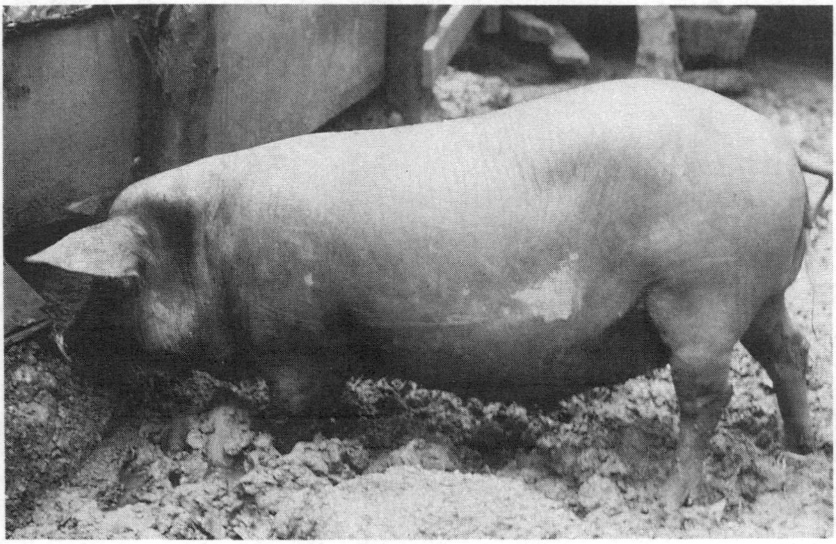

Fig 20 *An indigenous pig from Costa Rica*

The Large White (Yorkshire) (Figure 21)

The Large White breed was first developed in Yorkshire, England in the middle of the nineteenth century, and has since become a very popular breed throughout the world. It is a fast growing, strong-framed pig with good length and is renowned for its strength of leg. Females are prolific, good mothers and adapt well to confinement conditions.

The breed is widely distributed throughout the tropics, and is used extensively for cross-breeding. In Africa, for instance, the Large White X Landrace female is the most popular cross for commercial production. It is also used as grandparent stock in some of the main hybrids produced in Europe. In common with the Landrace, unless provided with adequate shade or wallows, the white skin renders it particularly susceptible to sunburn under tropical conditions. On the other hand, the white hair and white skin render the skin on the carcass more acceptable to the consumer than that from the coloured breeds.

The Landrace (Figure 22)

This breed originates from Scandinavia and is characterised by its forward-pointing lop ears. It was specifically developed for the bacon trade and typically possesses a long, smooth body with light shoulders and well-developed hams. It is a prolific breeder with excellent mothering qualities and produces lean, fast-growing progeny. The Landrace has a higher level of susceptibility to stress than some other breeds. Although not as numerous as the Large White, it is also common throughout the tropics, and is highly favoured for cross-breeding purposes.

The Duroc (Figure 23)

This breed has been developed in the USA, although there have been suggestions that the British Tamworth breed was involved in the original stock. It is characterised by its deep red or rusty colour. The Duroc is a fast-growing, large breed which has been selected specifically for overall muscle and meat production. One important characteristic is its ability to grow to heavier weights without depositing too much fat. In the female, litter-size and mothering ability are only average. It is claimed that the Duroc possesses a higher proportion of marbling fat in the meat.

An outstanding trait in the Duroc is its hardiness and resistance to stress, which results in lower levels of mortality. In tropical zones this is an important consideration, and the breed is consequently increasing in popularity. In commercial production in parts of Africa, it is frequently used as a terminal sire on white cross-bred females.

Fig 21 *A Large White boar*

Fig 22 *A Landrace boar*

Fig 23 *A Duroc sow*

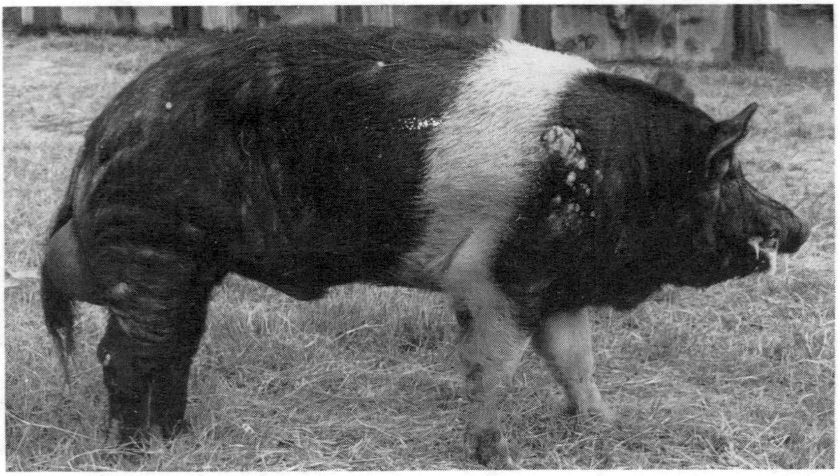

Fig 24 *A Hampshire boar*

The Hampshire (Figure 24)

The Hampshire is a medium-sized, black pig with a distinct white saddle which encircles the forequarter. Originally a native of England, it has been developed as a modern breed in the USA. Hampshire sows are prolific, good mothers and possess above-average milking ability. They are also better able to cope with more extensive conditions than white breeds. It is a meaty, well-muscled breed which shows good efficiency of food conversion. The Hampshire is very popular in cross-breeding programmes, both in order to produce a cross-bred female and as a terminal sire.

The Berkshire (Figure 25)

Although the Berkshire breed is on the decline on a worldwide basis, it remains popular in cross-breeding programmes in parts of the tropics. It is a smaller, early-maturing pig which was first developed in England for the pork trade. It has a black coat with characteristic white feet and nose. In the tropics it has proved very hardy, and crosses well with indigenous stock. In certain areas, for instance in Burma, it is prized for its fat content, as pig fat is used extensively for cooking purposes.

Other breeds of interest

The Chester White

This breed was developed in America and has spread particularly into Central and South America. Females are highly prolific, but growth rate tends to be slow, and carcasses are shorter and fatter than average.

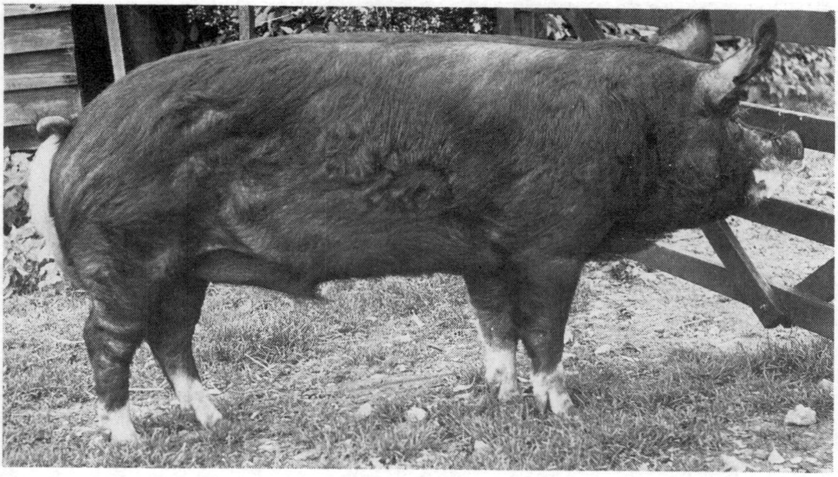

Fig 25 *A Berkshire boar*

The Large Black

A hardy British breed, sows are very good mothers under extensive conditions. However, growth rates tend to be slow and carcasses are relatively fat. Although now rare as a purebred, there is evidence of the influence of the Large Black throughout the tropics.

The Middle White

An early-maturing breed similar to the Berkshire, it originated in England. It was used extensively for cross-breeding in South East Asia but is also now rare.

The Pietrain

Of Belgian origin, this is a very lean and meaty pig, and is widely used in the production of modern hybrids. Introductions to the tropics have largely been unsuccessful due to the high level of susceptibility to stress. As an example, all the Pietrain pigs imported into Zimbabwe (then Rhodesia) in the 1960s died from heart failure as a result of one stress or another.

The Poland China

This breed of pig was one of the earliest to be developed in the USA. It tends to be large and fat but has been widely used in Central and South America. Improved strains of the breed are being developed.

The Tamworth

Characterised by its red colour, this is one of the oldest breeds of pig in

England. The breed is exceptionally hardy, but is relatively slow-maturing. In the past it has been very popular for cross-breeding purposes in tropical regions.

The British Saddleback

This is another hardy British breed, formed from the combination of the Essex and Wessex Saddleback breeds, with good milk production and mothering ability. It is named after its distinctive markings of black coloration with a white saddle. In the UK it has gained a new lease of life for the introduction of hardiness and mothering ability into hybrids used in outdoor production systems.

Genotype–environment interaction

From the point of view of genetic improvement of pigs in the tropics it is important to realise that pigs selected for their superior performance in one environment will not necessarily be superior in a completely different environment. This is known as genotype–environment interaction. An example of this type of interaction is that restricted feeding regimes will have a greater effect in reducing back fat thickness in genetically fat pigs than in genetically thin pigs.

Fig 26 *Exotic/indigenous cross-breds – better equipped to survive extensive conditions than the pure-bred exotic*

When selecting genotypes to help improve pig performance in tropical environments, it is crucial to understand the fact that such interactions will occur. If genotypes are selected under intensive conditions in a temperate environment and then transferred to the tropics, the environment must be modified, by means of housing, feeding and management, to suit that genotype. In extreme cases, genotypes selected under intensive, temperate conditions have been expected to perform under tropical scavenging conditions. In such situations the animals have difficulty surviving, let alone growing and reproducing, and local or cross-bred genotypes will be far superior under these conditions. (See Figures 17–20 and 26).

Methods of genetic improvement

The farmer has two main ways in which he can attempt to raise the performance of his pigs. He can either improve their environment or try to change their genetic make-up in order to increase their genetic potential. The various traits or characteristics of a pig are genetically controlled and inherited through genes, which contain the basic hereditary material. These genes can be manipulated to achieve genetic improvement by either increasing the frequency of favourable genes or combinations of genes by selection, or by introducing new genes into the population by crossbreeding with other breeds or strains.

Selection for improvement

Genetic traits can be divided into simple traits governed by a single pair of genes, such as shape of the ears or coat colour, or complex traits, controlled by many genes, which include the performance traits such as growth rate, food-conversion efficiency and carcass quality. With simple traits genes are normally dominant or recessive. If present in the heterozygote (i.e. a mixture of dominant and recessive genes), the dominant gene suppresses the expression of the recessive gene. Recessive traits will thus only appear when two recessive genes come together in the homozygous form (Figure 27). This means that the occurrence of a trait in a breeding programme can be predicted. As an example, if a recessive trait was desirable (e.g. prick ears in Figure 27) then only prick-eared animals would be used as parents. This pattern of inheritance was first discovered by an Austrian monk called Mendel, in his classic work with green and yellow peas, and is therefore known as Mendelian inheritance.

In the case of complex traits, the situation is entirely different. If we take growth rate as an example, within a given environment the individuals which possess a greater frequency of genes favouring growth rate

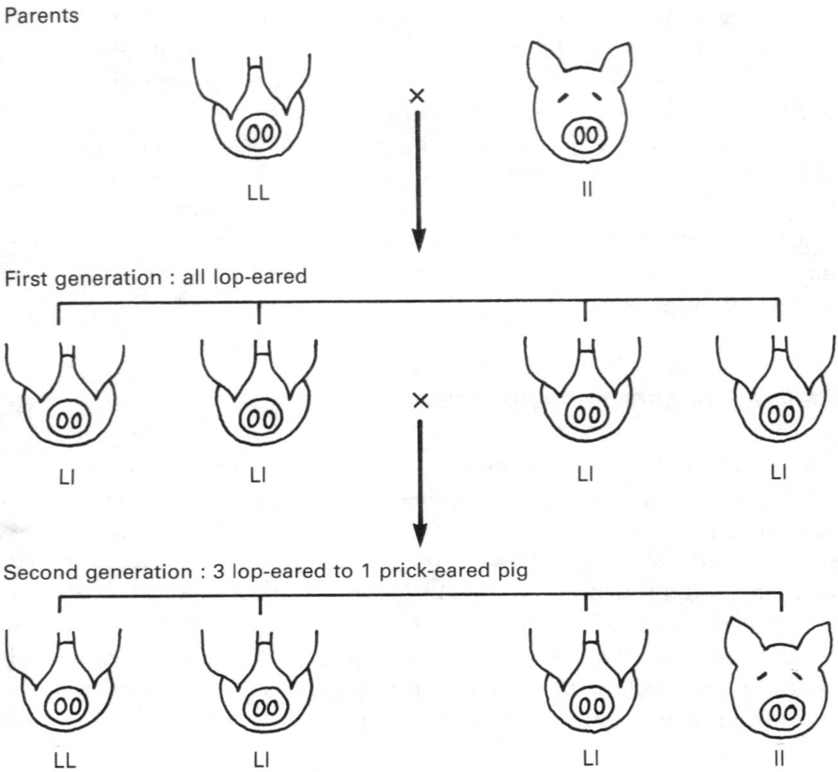

Fig 27 *The inheritance of simple traits: where lop ears (LL) are dominant over prick ears (ll)*

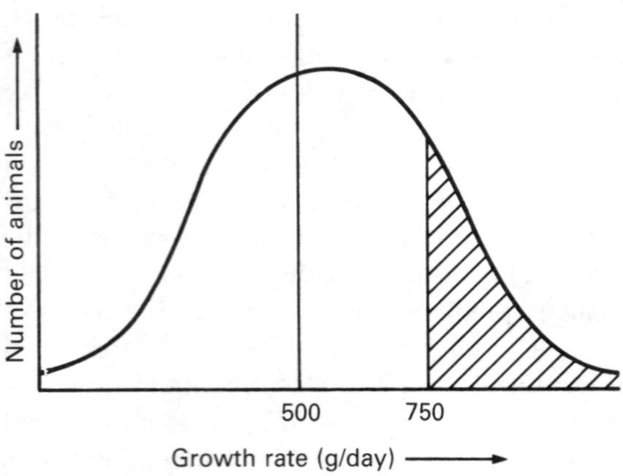

Fig 28 *With a complex trait such as growth rate, only superior animals within a population (shaded) are selected as parents*

will exhibit superior growth rates compared to the rest of the population. Thus, if only animals with superior growth rates are selected as parents, this will increase the frequency of the favourable genes in the next generation. This is illustrated in Figure 28, where a population of pigs has been plotted according to growth rate, and only the animals growing faster than 750 g/day would be used as parents.

The major factors which affect the efficiency and genetic progress in a selection programme are as follows.

Definition of objectives

It is of paramount importance that selection objectives are clearly defined before a breeding programme is embarked upon, and that they are not subject to constant change. This is particularly important in tropical conditions, where traits such as adaptability, coat colour and ability to produce on low-quality diets may be more critical than growth rates or carcass characteristics. Priorities may be completely reversed, e.g. fat pigs are preferred to thin pigs in some situations. The number of traits in a breeding programme must be kept to a minimum, as the more traits are selected for simultaneously, the slower the progress for each trait.

Selection differential

This is a measure of the superiority of the selected parents over the mean of the population from which they are derived. In Figure 28, for example, it represents the difference between the mean of animals in the shaded area (820 g/day) and the population mean (500 g). The bigger the differential, the greater the genetic progress. Clearly, the larger the variation in a heritable trait in a population, that is, the flatter the curve in Figure 28, the greater the scope for a bigger selection differential.

Heritability

Heritability is a measure of the proportion of the superiority of the parents above their contemporaries which on average is passed on to the offspring. More precisely, the heritability indicates the proportion of the total phenotypic variance that is due to additive genetic effects. Heritability estimates range from zero to 100 per cent. A zero heritability implies that the trait is not heritable because there is no genetic variation. Conversely, a heritability of 100 per cent indicates that the trait is totally heritable whereby differences in environments between animals will not affect the phenotypic variation of such a trait. In general in the pig, reproductive traits tend to be of low heritability, growth traits of medium heritability and carcass traits of relatively high heritability (Table 4.3). Genetic progress will always be greater when selecting for traits of higher heritability.

Table 4.3 Some heritability estimates for important traits in the pig

	Heritability %
Number piglets born per litter	15
Litter-size at weaning	7
Mean weaning weight	9
Daily weight gain	21–41
Feed-conversion efficiency	20–49
Dressing-out percentage	26–40
Mean backfat	43–74
Carcass length	40–87

Generation interval

This is defined as the average age of the parents when their offspring are born, and represents the time interval between generations. The shorter the generation interval, the more rapid the genetic progress, and if young boars are mated with gilts and replaced by selected progeny after one litter, the generation turnover can be as short as one year. Even in the normal situation where it is assumed that progeny born in first-to-fifth litters are equally likely to be chosen as replacements, pigs with an average generation interval of 2–2·5 years still have a great advantage over other domestic meat-producing species such as sheep (3–4 years) and cattle (4–5 years).

Accuracy of measurement of traits

The success of a selection programme is entirely dependent in the first instance on the accuracy of the records that are used. In some cases important traits are easy to measure, e.g. liveweight gain, but in others it is more difficult to be accurate, e.g. feed-conversion efficiencies and carcass measurements. Before embarking on a selection programme, it is essential to ascertain that the traits involved can be accurately measured.

Estimating the response to selection

Once the values for heritability (h^2) and selection differential (SD) and generation interval (GI) have been determined, the genetic gain per year can be estimated according to the formula:

$$\text{Genetic gain per year} = \frac{h^2 \times \text{SD}}{\text{GI}}$$

Testing procedures

Progeny-testing

As the boar has a relatively large influence on the characteristics which the next generation of a pig herd will inherit, testing systems have tended to concentrate on improvement of boars. Boar progeny-testing systems have been in operation throughout the world for a long time, and are based on measuring the relative merit of a boar's progeny from several sows. This obviously gives a true indication of what a boar may be able to contribute towards the genetic improvement of a herd. However, in addition to being very expensive, progeny-testing takes a long time in order to accumulate the data required to evaluate a boar, which makes him relatively old before his potential is known. This reduces his useful life. As a consequence, as a routine system, progeny-testing has largely been superseded for traits of higher heritability by the performance test. Nevertheless progeny-testing is useful for assessing traits which do not lend themselves to performance-testing, e.g. sex-associated and slaughter traits.

Performance-testing

The basis of a performance test is that an animal's own performance is taken as a measure of its genetic merit and with traits of high heritability used as a guide to how its progeny will perform. Thus the better individuals are selected from within a group of contemporary animals that have been treated similarly. The value and accuracy of a performance test can always be checked by running a subsequent progeny test and seeing if the results agree with the merit order of the performance test. Performance tests can be carried out at central performance-test stations, where the environment is standard for all animals tested. If facilities are adequate, testing can be carried out on-farm for within-herd comparisons.

The traits and selection criteria which are used in a test will obviously vary between countries according to their relative importance. They will also vary within that country depending on the use for which the pig is required. The criteria for selection of a boar which is to be used for generating gilts for commercial breeding, for example, will be different from those of a boar which is to be used as a terminal sire for the production of slaughter stock. Nevertheless, various combinations of growth rate, food-conversion efficiency and back-fat thickness are the traits which generally form the basis for selection.

Selection methods

Selection methods also differ and the two most commonly used are: independent culling levels and the selection index.

Independent culling levels

In this method, a level of performance is set for each trait, and if a pig fails to reach the desired standard in any trait it is automatically culled. It can be likened to an examination system where if you fail any subject, you have failed the total examination. A major weakness of this technique is that if a pig has outstanding qualities in some traits, say growth rate and food conversion efficiency, and just fails to reach the standard on conformation, it is culled. The genes for the outstanding traits are therefore lost. This method is the main system used for judging merit in pedigree breeding schemes.

The selection index

In the index method, the traits to be selected for are combined for one animal into a total score or index. Each trait is normally weighted according to its economic value and heritability, so that the highest-index animal should yield the highest financial return. These economic weightings in the index can be adjusted as economic circumstances change. The advantage of the index method is that exceptional performance in one trait can balance out a weakness in another, and if two traits are correlated so that improvement in one leads to a simultaneous decline in another required trait, this can be allowed for in the weighting.

Improvement programmes

As an example of how testing and selection procedures can be integrated into a genetic improvement programme, the national pig improvement programme in Zimbabwe is described. This programme is based on a pyramidal structure (Figure 29). A small number of registered nucleus

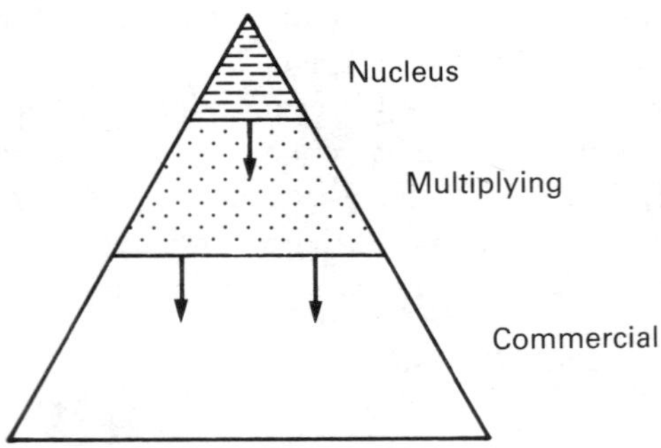

Fig 29 *The pyramidal structure of breed improvement in Zimbabwe*

Fig 30 *The central performance-test station in Zimbabwe*

Fig 31 *Young boars on performance test*

breeders assessed to be the best herds genetically in the country, send their pigs to a central performance-test station (Figure 30). Approved boars and gilts pass out to multiplication herds, assessed to be the next rung down in terms of genetic merit, where their progeny is performance-tested on the farm. Animals which pass this test are then available to commercial producers, thus ensuring a constant supply of quality replacement stock to commercial herds.

At the central test station, pigs are housed in pens of two and fed individually (see Figure 31), and their performance is measured from 35

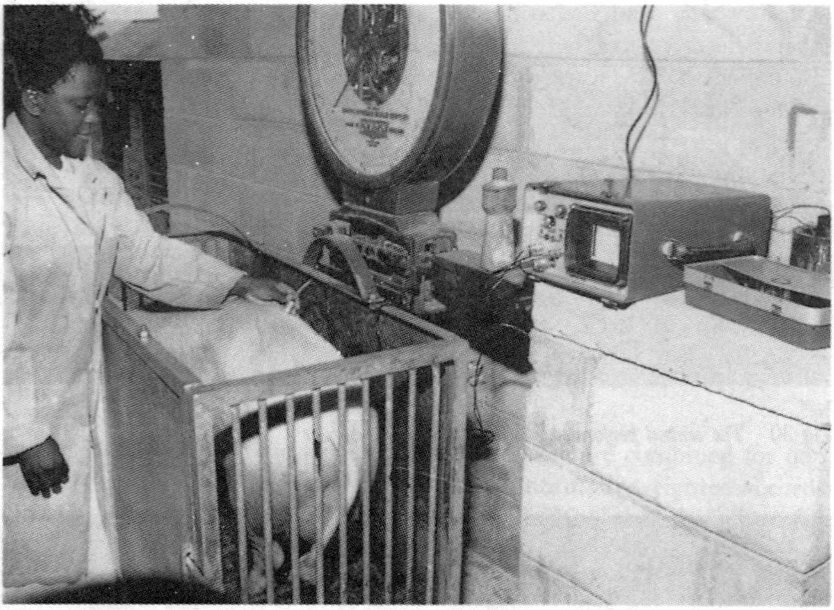

Fig 32 *Measuring back-fat thickness in a performance-test boar by an ultrasonic machine*

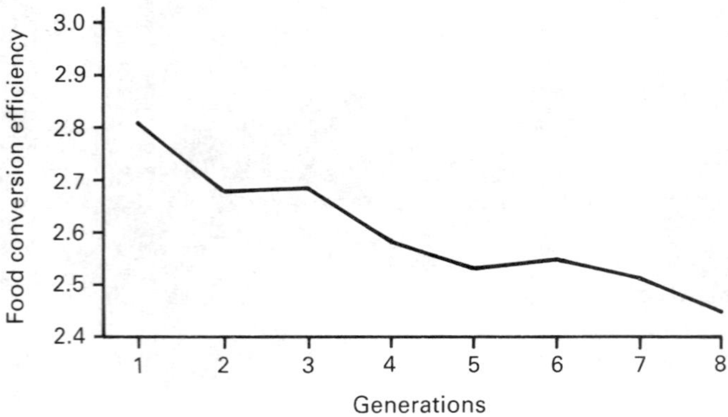

Fig 33 *The improvement in food-conversion efficiency as a result of central performance-testing in a control Landrace herd for eight generations in Zimbabwe*

to 86 kg body weight. At the end of the test, the food conversion efficiency is calculated, and the fat thickness is measured by an ultrasonic machine (Figure 32). Pigs are assessed on the basis of an index derived from the relative economic values of food conversion efficiency and fat thickness. Animals which do not pass the index are culled. In addition, animals with any abnormal sexual characteristics, blind or insufficient teats, or other defects such as weak legs or undesirable body conformation are also culled.

On-farm performance tests are then carried out on the sons and daughters of approved centrally-tested sires. Pigs are inspected for any weaknesses and, if they attain certain performance standards in terms of growth and fat thickness, they are made available for sale.

The potential value of regular performance-testing can be seen from the results of a control Landrace herd, in which both boars and replacement gilts were selected on the above scheme over eight generations (Figure 33). The improvement in feed-conversion efficiency represents a considerable saving in costs.

Practical cross-breeding

In the previous section, methods of pig improvement have been based on selection within breeds. There is, however, another means by which the producer can attempt to genetically improve the performance of his pigs and this involves cross-breeding, that is, the mating of individuals from two separate breeds or strains. The two major advantages which may accrue from cross-breeding are:

- exploitation of the phenomenon of 'heterosis', which occurs when two breeds, which are genetically different, are crossed. If the cross-bred individual shows an improvement in performance above the mean of both parents it is said to exhibit hybrid-vigour.
- complementarity, or the ability to combine in one or more individuals the desirable characteristics of two or more existing breeds.

The degree of improvement which may be expected from heterosis is shown in Table 4.4.

Table 4.4 Estimated percentage levels of heterosis for some traits of economic importance in pigs

	+% of mid-parent value
Litter-size at birth	2–8
Litter-size at weaning	5–11
Total litter weight at weaning	10–12
Post-weaning growth	6–8
Carcass traits	0–5

The higher levels of heterosis tend to occur in traits which are of lower heritability, e.g. reproductive traits. The degree of heterosis will always tend to be greatest when there is most genetic divergence between a tropical indigenous and an imported exotic breed. For this reason, the level of heterosis will always be highest in the first (F_1) cross, and decrease in subsequent crosses as the genetic differences decline. However, the overall economic gain which can be obtained from heterosis effects can be cumulative within a cross-breeding system. Thus for litter-size at weaning, for example, if the first cross dam gives a heterosis improvement of 11 per cent, and there is a further 6 per cent to be gained from her progeny's performance, the cumulative effect is 17 per cent, or more than one additional pig weaned per litter. There are a number of different cross-breeding strategics which can be used, all of which harness different amounts of heterosis.

Of these, the two-breed crisscross system is probably the most appropriate for the small-scale producer in the tropics (Figure 34). If a third pure breed is available this can be extended to a triple crossing system using a sire selected from each of three breeds in rotation. Maximum heterosis is obtained when the genetic diversity between the parent breeds is large, and there is therefore considerable potential for crosses between exotic and indigenous tropical breeds. The best use of such crosses is likely to be in semi-intensive systems of production (see Chapter 2), where the hardiness and foraging ability of the indigenous pig will be complemented by the growth and improved performance of the exotic and large numbers of cross-bred animals can be found on these systems (see Figure 26). The two main problems are the maintenance of exotic boars for generating the cross-breed, as they often fail to survive due to disease and management hazards, and the lack of breeding control resulting in indiscriminate interbreeding. One possible solution under such conditions is the use of institutionally managed boar-holding centres, such as those in some parts of northern Nigeria, where indigenous sows from small-scale farmers are kept to be mated with pure-bred exotic boars. Alternatively, private farmers with intensive production units could be encouraged to produce cross-bred progeny for sale to small farmers. Similar methods could be used for the generation of exotic × indigenous boars for use on indigenous sows in small-scale production systems.

The benefits of heterosis in female reproductive traits can also be utilised in boars. Cross-bred boars have been shown to have greater libido, larger testes and higher sperm counts than their pure-bred contemporaries, leading to more reliable breeding and improved conception rates. This enhanced reproductive fitness is obviously likely to be a major advantage in extensive and semi-intensive pig production systems in developing countries.

The harnessing of heterosis is maximised in the genetic improvement schemes run by the private breeding companies in the developed world. Each company maintains a network of super-nucleus and nucleus herds

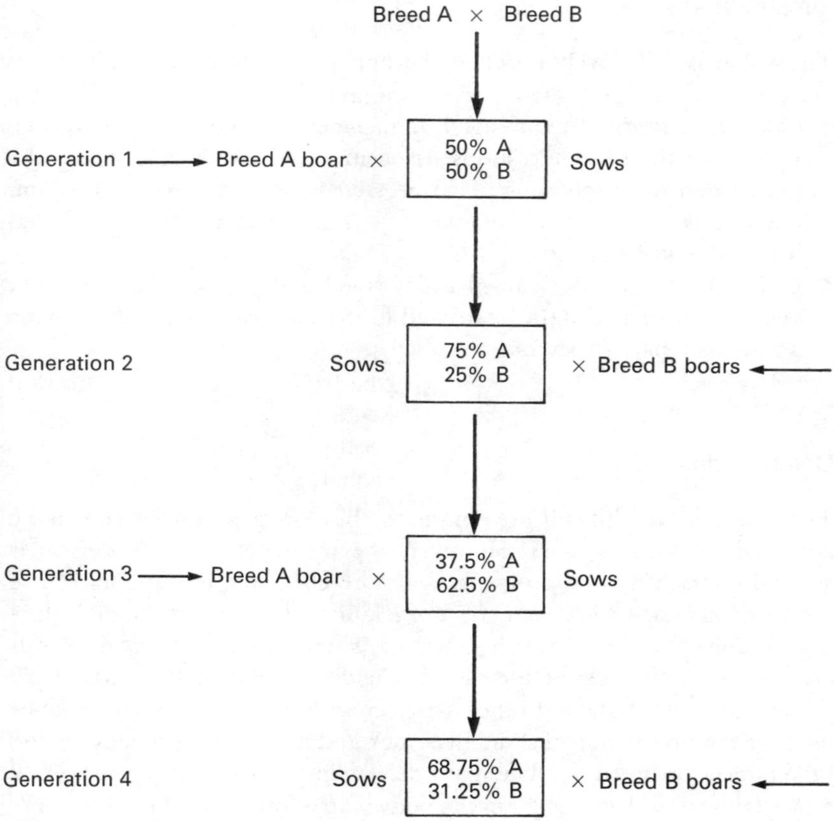

Fig 34 *Criss-crossing with two breeds of pig*

in which superior genes are concentrated. These herds are subjected to strong selection pressure with rigid and comprehensive testing of stock for important traits. Selected purebred stock from nucleus herds is then used to populate multiplier herds where improved lines and breeds are crossed. The resultant first-cross or hybrid gilts are sold to commercial producers.

Inherited defects

There are many anatomical defects in pigs which can affect performance. The majority of these are genetic in origin and therefore likely to be inherited from generation to generation. These defects may be simple (i.e. controlled by one pair of genes) or complex (controlled by several pairs of genes).

Simple defects

These can be effectively eliminated from a herd by culling. The main defects in this category are:

- *Congenital tremors* In this condition, piglets are born with rhythmic tremors of the head and limbs. It occurs at a frequency of about 0·1 per cent in the population. The recessive gene is carried by the dam, and in carriers 50 per cent of male piglets are affected. Dams of affected litters should be culled.
- *Club foot* This defect causes a deformed and swollen foot. It has a very low incidence and is restricted to the Landrace breed. Both male and female parents should be culled.

Complex defects

These are more difficult to eliminate, but greatest response will be achieved by culling boars. However, the incidence of these defects is generally relatively low, so it may be more economic in the long run to keep a good boar which is a carrier of a known defect rather than replace it with a genetically inferior boar which is not a carrier. If good records are available, the cost of the defect can be set against the gain in pig performance attributable to the carrier boar. Such a boar should only be used for the production of slaughter stock and none of his progeny should be retained for breeding. The main defects in this category are:

- *Scrotal hernia* A weakness in the body wall allows part of the intestines to pass out into the scrotum. It occurs at a frequency of around 0·7 per cent and has an estimated heritability of 15 per cent.
- *Umbilical hernia* A similar condition which occurs at the site of the umbilicus. Found at a frequency of around 0·6 per cent.
- *Imperforate anus* The incidence of this condition in piglets is around 0·35 per cent. Mortality in male piglets is always 100 per cent, but often around 50 per cent of females survive as the faeces are voided via the vagina.
- *Splaylegs* Piglets are born with either the front or hind legs splayed, sometimes both, and are unable to stand. The incidence is around 1·5 per cent, and the condition can be worsened by nutritional deficiencies. If piglets survive for three days, the condition often tends to disappear.
- *Hermaphroditism* In this condition, females tend to exhibit male characteristics. Incidence is estimated at 0·07 per cent.
- *Cryptorchidism* Also known as a 'rig' pig, one testicle in the male fails to descend into the scrotal sac. Found at a frequency of 0·22 per cent.
- *Female genital defects, including 'inverted nipples'.* These occur at a frequency of around 0·15%

Stress susceptibility – the 'halothane gene'

Throughout the world, many breeds of exotic pigs exhibit a condition known as porcine stress syndrome (PSS). This renders the animal particularly susceptible to stress and when stress is caused, for example by mixing, transport, exercise, fighting or mating, the pig will often suffer an irreversible rise in temperature followed by sudden death. Slaughtered pigs will also show a high incidence of pale soft exudative (PSE) meat which reduces the attractiveness of the meat to the consumer.

It has now been discovered that, surprisingly, the condition is controlled by a single pair of genes which behave in a Mendelian fashion. That is, the pig exhibits PSS when two recessive genes come together according to Mendelian laws, as was demonstrated for the prick-ear trait in Figure 27.

The ability to monitor the presence of this recessive gene in a population has been facilitated by the further discovery that when subjected to halothane gas anaesthesia, PSS pigs which are homozygous for the recessive gene become rigid and hyperthermic, whereas stress-resistant pigs which are either heterozygous or do not possess the recessive gene at all become relaxed. Thus it has become known as the 'halothane gene'. In different breeds, the incidence of a positive reaction to halothane anaesthesia has varied from nil in the Large White and Duroc to over 80 per cent in the Belgian Landrace and Dutch Pietrain.

If it is required to cull pigs which are recessive for the halothane gene, they can easily be identified by using halothane anaesthesia. On the other hand, the heterozygous condition confers some advantages, such as good muscling ability and leanness, on the pig. The advantages of the heterozygote can be exploited by using positive halothane reactor boars as terminal sires on female lines which are free from the gene.

Artificial insemination

Artificial insemination (AI) involves the collection of semen from a boar and then the introduction of semen into a sow or gilt at a later stage by means of a catheter. This contrasts with natural service where a boar mounts a sow and introduces his semen.

Benefits

The major advantage of AI is that it allows for the wider use and distribution of boars of high genetic merit. The ejaculate from one boar can be extended to inseminate up to 25 sows. Recent advances in methods of boar semen storage make it possible for developing countries in the tropics to import the very top genetic stock from developed countries (e.g. in the UK, only the top 5 per cent of boars performance-tested by the Meat and Livestock Commission are eligible for entry to AI studs). This calibre of genetic material would not otherwise be available to developing counries.

Other benefits of AI are:

- it overcomes the need to purchase, house and feed a boar. This is particularly pertinent to the small-scale producer who cannot justify keeping a boar for a small number of sows, and who cannot afford a boar of good quality. The effective use of AI is especially relevant where small-scale producers are involved in group or co-operative pig development schemes, and their units can be serviced from a central boar-holding centre.
- it prevents the transmission of disease from farm to farm by the sale and purchase of boars and on-farm reproductive diseases cannot be spread by boar-to-sow contact.
- it overcomes the practical problems of differences in size of males and females. On occasions, this problem can severely limit the use of boars of high calibre.
- it reduces the risk to stockmen of handling boars for natural service.

Techniques

(a) Semen collection

Although various artificial vaginas and electro-ejaculators are available, they are not necessary for successful semen collection. Boars can be easily trained to mount a dummy sow device or an oestrous sow, and firm pressure on the penis by a gloved hand causes ejaculation to occur. The first low-sperm fraction of the ejaculate should be discarded, and then the second sperm-rich fraction can be collected through a filter funnel, which removes the gelatinous fraction, into a warmed (30°C) bottle.

A drop of semen can be observed under a microscope to check its fertility characteristics and, if desired, the semen can then be diluted. A number of diluents and extenders are available, and the individual doses are normally stored in 50 ml plastic bottles for up to three days at 15–20°C. The number of spermatozoa used under commercial conditions for one insemination normally varies from 1×10^9 to 3×10^9.

(b) Insemination

This involves the insertion of a rubber spiral catheter (Figure 35) into the sow's vagina, and then rotating it in an anti-clockwise direction (Figure 36(a)) until the tip locks into the cervix. The bottle containing the semen dose can then be attached to the other end of the catheter and the semen runs in under gravity or slight pressure (Figure 36(b)). The insemination process may take up to 15 minutes.

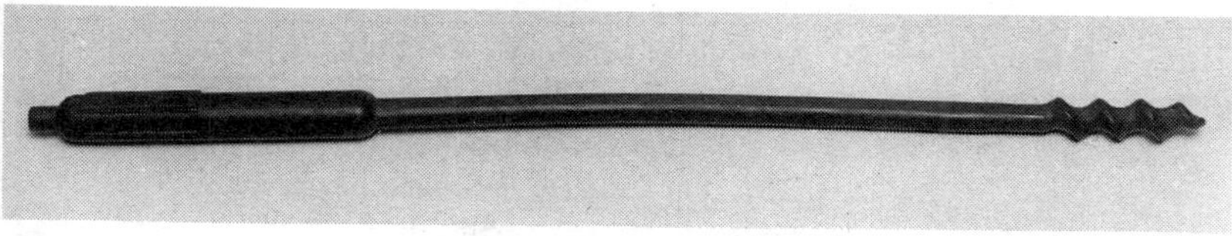

Fig 35 *A rubber catheter used for artificial insemination*

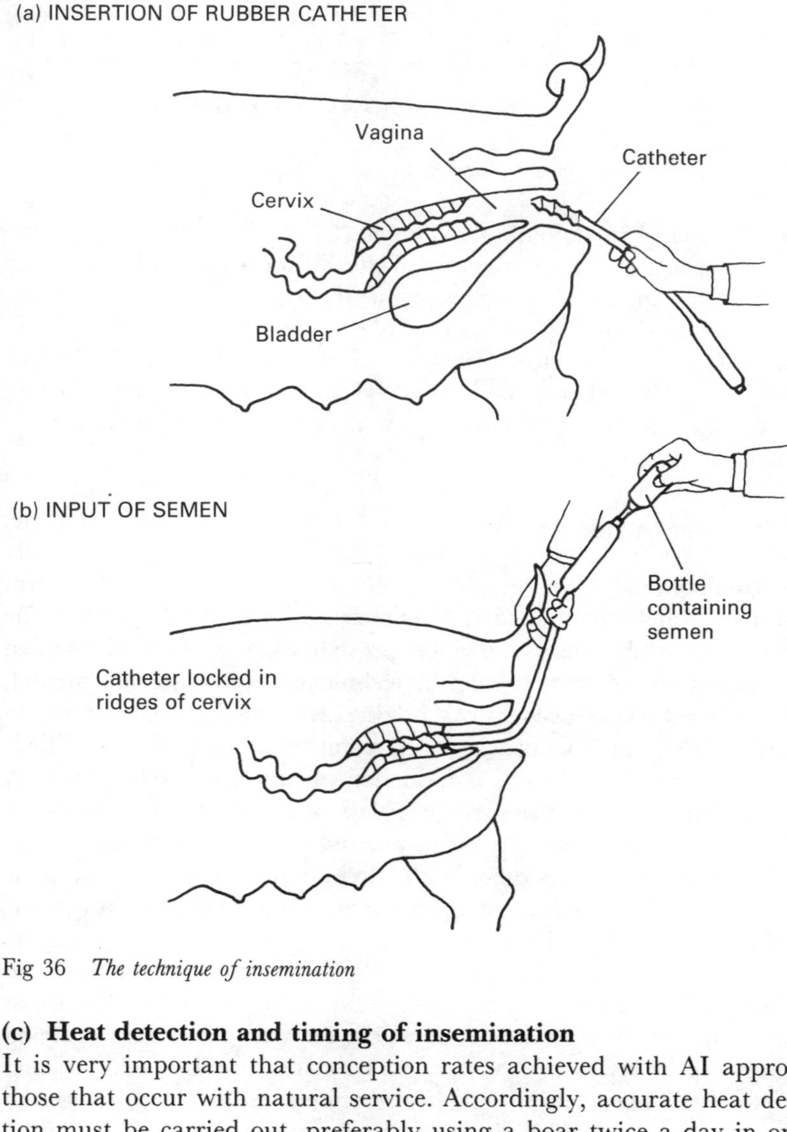

Fig 36 *The technique of insemination*

(c) Heat detection and timing of insemination

It is very important that conception rates achieved with AI approach those that occur with natural service. Accordingly, accurate heat detection must be carried out, preferably using a boar twice a day in order

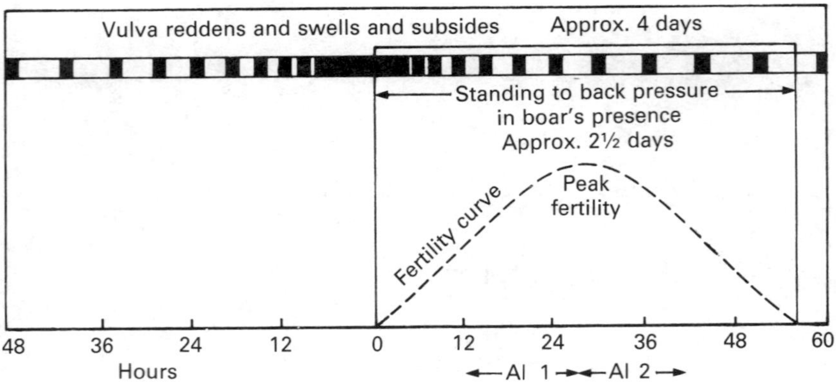

Fig 37 *Recommended times for insemination*

that the timing of insemination is correct. To overcome inaccuracies in the detection of the start of oestrus and the natural variations in time of ovulation, two inseminations approximately 12 hours apart are recommended (see Figure 37). More recently, devices have been developed which measure the electrical resistance of the vaginal mucosa. As this varies in relation to hormonal events, it can be used to predict more accurately the timing of ovulation, and hence the optimum timing of insemination.

System of semen storage

(a) Frozen semen

In contrast to bull semen, the nature of boar semen renders it susceptible to damage by the freezing and thawing procedures, and as a consequence only relatively recently have successful techniques for freezing boar semen been developed. Frozen semen can be obtained in either pellet form or in straws, and acceptable conception rates obtained. This has provided a major breakthrough for the introduction of superior genetic material into developing pig industries. Nevertheless, strict attention to detail in the handling of the semen has to be adhered to if successful results are to be achieved. Also boars differ in the 'freezability' of their sperm, and the semen of individual boars must therefore be screened before it is used for the production of frozen semen.

(b) 'Long-life' semen

Extenders have now been developed which allow for the storage of fresh semen for up to seven days without any marked loss in fertility. This allows for fresh semen to be sent by air around the world and used to impregnate sows successfully in the country of destination.

5 Principles of feeding and feed resources in the tropics

Introduction

In order to survive, maintain itself, grow and reproduce, the pig requires a supply of essential nutrients. Nutrients are present in various feedstuffs and food which is eaten by the pig is broken down into its component parts in the digestive tract (see Chapter 2). These parts or basic nutrients pass into the bloodstream of the animal and are then used for various processes according to the biological needs of the pig. The major groups of essential nutrients are energy, protein, minerals, vitamins and water.

Energy

Definition

Apart from water, sources of energy are the most important food requirements of the pig and will most rapidly influence its survival if withdrawn. Energy can be defined as the capacity to do work and occurs in various interconvertible forms such as chemical, thermal or radiant energy. It is normally measured in heat units, traditionally the calorie, but in modern livestock nutrition, the megajoule (MJ) is the most commonly-used unit (where 1 MJ = 0·239 MCals).

Systems of measurement

The energy value of a feed is measured in the first instance as the Gross Energy (GE), which is the amount of heat produced when the food is completely burnt in oxygen. Because not all this gross energy is available to the pig (that is, it is not digested), digestible energy is widely used to describe the energy content of pig feeds, where:

digestible energy (DE) = GE − energy loss in faeces.

In some instances, Metabolisable Energy (ME), which allows also for energy losses in the urine and combustible gases, is used as a measure for energy. However, as ME is difficult to measure precisely and because it approximates fairly closely to the DE value (ME = 0·96 × DE), DE values are in general use.

An alternative system of assessing the energy value of pig feeds is the Total Digestible Nutrient (TDN) system, which is widely used in the USA. TDN is calculated as the sum of the digestible components of the diet, namely:

TDN = digestible crude protein
+ digestible crude fibre
+ nitrogen-free extract
+ (ether extract × 2·25).

One kg of TDN is equivalent to 18·49 MJ DE.

Requirements for maintenance

A pig is in a state of maintenance when its body composition remains constant, when it does not perform any work or give rise to any product (e.g. milk). The energy expended for the maintenance of an animal is converted into heat, which leaves the body. The quantity of this heat gives a measure of the pig's basal or fasting metabolism, and a direct estimate of the quantity of heat energy that the pig requires from food for maintenance. Thus the total requirements of the pig for energy can be distinguished between those required for maintenance and those for production. In normal practical circumstances, an absolute maintenance situation, where the animal is in a state of equilibrium without gain or losses of fat or protein, is unlikely to occur. However, it is very relevant that, with the vagaries of feed supply in the tropics, adult pigs may spend periods of time close to a maintenance situation. Maintenance requirements are considerably affected by environment, particularly temperature, as both the retention and loss of heat require an energy input.

Another aspect of the requirements of energy for maintenance is that, although related to body size, they do not increase as a fixed proportion of body weight, but become proportionately less in relation to body weight as the pig grows. In fact, maintenance requirements are more closely related to the surface area of the pig and lie somewhere between the two. Metabolic bodyweight is defined as bodyweight to the power ¾ ($w^{0.75}$).

Requirements for production

Energy requirements for production are less dependent on the environment but are strongly influenced by the genotype of the pig, and this means that rations must be designed according to the potential performance of the animal. In addition, at any one time, energy requirements will depend to a large extent on the type of production. If it involves the laying down of fat, then the energy costs are high as fatty tissue contains only small amounts of water, whereas if it is protein deposition, which involves some 75 per cent of water, then the energy costs are relatively low. Similar considerations apply to estimating the energy requirements for pregnancy and lactation.

Of the total energy available to the pig, obviously only that which is surplus to requirements for maintenance is available for productive processes. This surplus is restricted by the pig's ability to eat, which declines in relative terms as the pig grows. Whereas a 15 kg pig can eat about four times the feed it needs to maintain itself, this reduces to three times or less at 90 kg liveweight.

Fibre

As described in Chapter 2, the enzymes in the pig's digestive tract cannot digest fibre which occurs to some extent in all plant material. However, the bacteria in the caecum can break down a small amount of fibre to fatty acids such as acetic, proprionic and lactic acid, which are then available as a source of energy. In general a high level of fibre in the diet will reduce the availability of other energy sources, particularly if the fibre is not ground or milled. Of more importance is that fibre in the diet promotes a healthy gut environment and ensures an efficient rate of passage of digesta.

Protein

Composition

Protein consists of some 20 basic units known as amino-acids. When protein is eaten by the pig, it is broken down to the amino-acids, which are then used for the building up of the essential organs of the body and skeletal muscle (lean tissue). Protein makes up some 15 to 20 per cent of the total bodyweight of the pig.

Essential amino-acids

Nine of the amino-acids required by the pig cannot be synthesised by its body, and they must therefore be supplied in the diet. If they are present in insufficient quantities, the pig will not grow and may not even survive. These are known as the essential amino-acids in contrast to those which can be manufactured in the body (Table 5.1) which are known as non-essential.

Table 5.1 The essential amino-acids for pigs, and the amounts of essential amino-acids in the 'ideal' protein for growing pigs

Essential amino-acids	Amounts of ideal requirement for growing pigs (g/kg protein)
Lysine	70
Methionine and cystine	35
Threonine	42
Tryptophan	10
Isoleucine	38
Leucine	70
Histidine	23
Phenylalanine and tyrosine	67
Valine	49

(Source: ARC, 1981)

Ideal protein

The abundance of the various essential amino-acids in the diet is also of importance in relation to the priority of their need by the pig. Valine and isoleucine, for example, are in relatively plentiful supply in foodstuffs, whereas lysine and methionine are in shorter supply but are required in larger quantities by the pig. When the supply of essential amino-acids in the diet approximates to the ratio that is required in the body of the pig, it is close to an 'ideal' protein from the pig's point of view. The Agricultural Research Council in the UK has listed the amino-acid composition of an 'ideal' protein source (Table 5.1).

Biological value

Protein quality is assessed on how closely the protein approaches the ideal balance of essential amino-acids and this is known as the biological value of the protein. In practical terms, the total utilisable protein in a feed depends on the first limiting essential amino-acid. In parts of the tropics, for example, where maize is often the staple ingredient of pig diets, lysine is usually the first limiting amino-acid. It is important that when it is fed to pigs, maize is complemented by a protein source with a relatively high lysine content, e.g. soyabean meal. Groundnut meal, which is another common protein source in the tropics, has a higher protein content but less lysine than soyabean meal and is therefore of lower value for the pig.

Digestibility

The amount of amino-acids which become available to the pig will also depend on the digestibility of the protein. The main factors which affect protein-digestibility are:

- *Heat damage* Digestibility of the protein will fall dramatically if the protein is heat-damaged, which can often occur in the preparation of protein meals. Specific amino-acids can become bound to other compounds on heating and therefore unavailable. A prime example is lysine in blood meal where the theoretical lysine content is relatively high, but the 'available' lysine is low.
- *Degree of grinding* If feeds are not ground, enzyme penetration and digestion is often prevented by indigestible plant material.
- *Poorly-digestible carbohydrates* If the proteins are contained within plant cells which are resistant to breakdown, the proteins cannot be released and will pass through the gut undigested.
- *Other anti-nutritive factors.*

Protein: energy ratios

Efficient utilisation of protein is dependent on the amount of energy available. Thus the amount of protein per unit of DE is more important than the absolute concentration of protein. The optimum ratio depends on:

- *The age of the pig* The optimum protein-to-energy ratio changes steadily as the pig grows, being high in the young animal and lower in the older pig, when protein requirements per kg liveweight are reduced.
- *Genotype* Exotic pigs, for example, have requirements for a higher protein-to-energy ratio than unimproved and indigenous pigs, because of a higher lean-to-fat ratio in their bodies (See Figure 6).

Maintenance, production and surplus

In a similar way to energy, the first call on available protein in the body is for maintenance. This supplies the need for tissue replacement and repair, manufacture of enzymes and hormones. In the growing animal this is usually less than one-third of the total requirement for protein. Additional protein is then needed for growth and the deposition of lean tissue, or for milk synthesis.

Protein which is surplus to these requirements, or amino-acids which are in excess and therefore cannot be utilised by the pig, are deaminated in the liver. The nitrogen is excreted in the urine and the remaining carbohydrate fraction is used as an energy source.

Minerals

Introduction

The growth and efficiency which should accrue from expensive inputs of protein and energy can be negated by small imbalances in or shortages of essential minerals. Defining the pig's requirement for minerals is complicated by the interaction that occurs between some minerals and the fact that not all the minerals in a foodstuff are available to pigs because they cannot be digested.

Requirements

Estimated requirements for minerals by the main categories of pig are shown in Appendix A. The essential minerals in the diet are as follows:

Calcium and phosphorus

Deficiency of either calcium or phosphorus results in inadequate bone calcification, and bones becoming brittle and easily fractured. This leads to rickets (in younger pigs) or osteomalacia (in mature pigs). If calcium or phosphorus are deficient in the diet of lactating sows, the sow will draw on supplies from her skeleton, resulting in lameness, rear-leg paralysis, and the 'Downer Sow Syndrome', when the sow becomes irreversibly crippled and unable to rise.

The calcium:phosphorus ratio in the diet is also important, and should not exceed 1·7:1 for pigs up to 20 kg liveweight; 2·0:1 for pigs from 20–55 kg; and 2·4:1 for pigs above 55 kg.

Magnesium

Deficiency will lead to poor growth, bowed legs and a stiff-legged gait known as the 'Stepping Syndrome'. Generally, however, there is adequate magnesium in the diet, especially in maize–soyabean meal diets.

Sodium and chlorine

These are treated together as it is common practice to add them to pig diets in the form of common salt ($NaCl$). Correct levels of salt are of prime importance in hot climates, as deficiency gives rise to poor appetite and low water consumption, leading to general ill thrift.

If water is short, toxicity can also be a problem, as this will occur at levels of 2 per cent of salt in the dietary dry matter. Toxicity is characterised by weakness, nervousness, convulsions, staggering and death.

In certain areas, particularly dry ones, water quality may be affected by too-high concentrations of salt. This increases the chances of toxicity and lowers water intake. In these areas, water may need to be treated before being offered to pigs.

Copper
Copper deficiency results in nutritional anaemia and leg weakness, and this is probably associated with the copper/iron interaction.

As copper toxicity occurs only at levels of 500 mg per kg DM in the diet, which is approximately 125 times estimated requirements, copper is frequently included well above minimum requirements in grower diets as a growth promotant.

However, care must be taken with young pigs as an inclusion rate of 300 mg/kg of diet can be toxic especially if zinc and iron levels are low.

Iron
Lack of iron results in anaemic pigs, with poor growth and a rough hair coat. It is a particular problem in piglets, because they are born with low iron reserves and there is very little supplied in the sow's milk. As a consequence, piglets reared intensively without access to soil require a supplementary source of iron, either given orally or by injection (see Chapter 8).

Zinc
A deficiency of zinc gives rise to the condition known as parakeratosis. This causes red raised areas on the skin which become crusty, and may eventually coalesce over the whole body. Other symptoms are poor growth and reduced appetite.

Manganese
Deficiency of manganese will lead to lameness in growing pigs and the birth of weak piglets from deficient sows. However, a deficiency is not common in normal pig diets.

Iodine
The main symptoms of an iodine deficiency is goitre and an enlarged thyroid. Sows will farrow weak, hairless pigs.

Selenium
Because of their interaction, selenium must be considered together with Vitamin E. A lack of one will affect the utilisation of the other. There are three main conditions resulting from deficiency, namely: mulberry heart disease, hepatosis dietetica and muscular dystrophy. Mulberry heart disease normally affects pigs from three weeks to four months of age, and the main symptom is sudden death due to cardiac failure. Post-mortem examinations will show lesions of the cardiac muscles. Hepatosis dietetica also occurs between three weeks and four months of age and is characterised by a generalised oedema and death, with the ears turning blue. With muscular dystrophy, pigs develop a staggering gait and are

unable to rise. It usually affects fast-growing pigs from 30 to 60 kg liveweight.

Vitamin E is broken down by oxidation if stored at high temperatures. It therefore tends to be unstable under tropical conditions and unless an antioxidant is added to the rations, a deficiency is likely to occur. This is particularly the case when selenium levels in the diet are low, as in the case when maize–soyabean meal combinations are used as food, or when the levels of unsaturated fats in the diet are relatively high.

Availability

Availability can firstly be affected by the source of the mineral. Phosphorus, for example, if supplied to the pig as phytate phosphorus, remains bound to the calcium component, and its availability is markedly reduced. Secondly, availability can be affected by the metabolic state of the animal. It has been shown that calcium and phosphorus availability will decrease with age and size of pig from 85 and 80 per cent at 5 kg bodyweight to 45 and 65 per cent respectively at 90 kg bodyweight.

Interactions

Interactions can play a major role in mineral utilisation. These can occur between minerals, between minerals and vitamins, and between minerals and other feed ingredients, where one ingredient can depress or enhance the utilisation of the other. Some important interactions are:

- Calcium, phosphorus and Vitamin D
- Vitamin E, selenium and fat levels
- Calcium and phosphorus influencing the metabolism of zinc
- Copper and zinc
- Copper and iron
- Iron, Vitamin E and selenium

Vitamins

Definition

Vitamins can be defined as organic compounds which function in small amounts (mg or μg) and are essential to the normal functioning of the animal body. They cannot be synthesised in adequate amounts by body tissues and when lacking, provoke deficiency diseases.

Requirements

Estimated requirements for vitamins by the main categories of pig are shown in Appendix A. Fourteen vitamins are normally considered to be required in the diet of pigs, of which four are fat-soluble and ten are water-soluble.

The fat-soluble vitamins are:

- *Vitamin A* Deficiency causes night blindness, birth of blind piglets, lacrimation of eyes and nasal discharges and rough and dry hair coat. It is also involved in reproduction, and lack of Vitamin A will lead to a decrease in ovarian and testicular size and increased embryo mortality.
- *Vitamin D* Through its relationship with calcium and phosphorus metabolism, deficiency can lead to rickets and osteomalacia. If pigs have access to sunlight, Vitamin D may be synthesised in adequate amounts in the skin.
- *Vitamin E* Interacts closely with selenium, and the main deficiency symptoms are listed under selenium. Deficiency will also lead to reproductive disturbances and reduced milk production.
- *Vitamin K* Deficiency of Vitamin K does not often occur, but when it does will cause a decrease in the clotting time of the blood, nasal bleeding and subcutaneous haemorrhages.

The water-soluble vitamins are:

- *Thiamin* Deficiency causes nervous disorders, cardiac lesions and high piglet mortality at birth.
- *Riboflavin* Typical symptoms of riboflavin deficiency are a rough hair coat, loss of hair and dermatitis. There has been renewed interest in the importance of riboflavin for sows. Deficiency in sows leads to resorption of the foetus, premature farrowing and anoestrus in gilts.
- *Niacin* Lack of niacin causes dermatitis of the skin, diarrhoea and anaemia.
- *Vitamin B6* Slight deficiency will cause vomiting and diarrhoea, but more severe deprivation leads to disordered movements, lacrimation and blindness.
- *Pantothenic acid* Deficiency causes scurfy skin and leads to the locomotion disturbance known as 'goose stepping'.
- *Biotin* Lack of biotin leads to a rough hair coat, cracks in the feet and lameness, and depressed reproductive efficiency.
- *Folic acid* Because a deficiency of folic acid causes lesions in the mouth and buccal cavity, this renders the pig more susceptible to infection.
- *Vitamin B12* This is an important vitamin in relation to general health, and a lack of it will cause diarrhoea, vomiting, rough coat, lack of co-ordination in the hind legs and reduced reproductive performance.
- *Choline* A deficiency of choline causes splay legs in baby pigs and

reduces piglet survival. It is also involved in reducing reproductive performance.

- *Vitamin C* Although not common, a lack of Vitamin C in the diet will lead to weakness and haemorrhaging throughout the body. Vitamin C is often fed in extra amounts to improve the immune response and resistance to infection.

Availability

The availability of a vitamin to the pig can vary and can be influenced by a number of factors. The B vitamin, niacin, for example, is present in relatively large quantities in cereal grains, but is bound in a form which makes it unavailable to the pig. Certain antagonists may also occur in the feed which interfere with utilisation of the vitamin.

Losses in storage

As a generalisation, it can be stated that the vitamin potency of feeds will tend to decline during storage. The rate of decline will depend on the conditions of storage, e.g. the intensity of light, ambient temperature, acidity, the presence or absence of interfering substances and the physical form of the diet.

Because storage conditions in tropical areas are often far from ideal, it is very important that the periods of storage of mixed foods are minimised. Moreover, a synthetic antioxidant (e.g. methylene blue or ethyoxyquin) should always be included when vitamin supplements are compounded, in order to increase the stability of the vitamins.

Interactions

As in the case of minerals, the function and effectiveness of a vitamin may be influenced by interaction with other vitamins, minerals or feedstuffs.

Water

With the exception of oxygen, water is the most important ingredient for the maintenance of life. Some 65 per cent of the body of a pig is comprised of water and it is involved in most of the chemical reactions which take place in the body.

Water is of particular importance in the tropics and hotter and drier parts of the world, where it is often in short supply, and the pig requires more water to enable it to maintain body temperature. Lack of water very quickly leads to a rise in body temperature and death. However, equally important is that sub-optimal amounts of water will have a major effect on food intake and pig performance, and this applies irrespective of the system of production. It is thus essential that, in tropical climates, pigs have access to a supply of clean fresh water at all times.

Anti-nutritional factors

In the tropics, plant proteins are by far the most common source of protein for pigs and a number of these are associated with anti-nutritional factors. These can be in the form of toxins or other substances which interfere with digestion and utilisation, and it is essential that these factors are taken into account when formulating rations. A prime example is soyabean which contains a trypsin inhibitor. This inhibits the action of trypsin in the pig's gut and can reduce the digestibility of protein so that only 30 per cent of it will be digested.

Voluntary food intake

The more food the pig eats, the more nutrients are likely to be available above maintenance requirements for productive processes. Recent research has shown that if appetite in young pigs can be increased above normal levels, levels of protein deposition and growth are markedly enhanced.

The pig will normally control its energy balance by regulating its voluntary feed intake. This is of particular relevance under tropical conditions where pigs often receive diets of lower nutrient, particularly

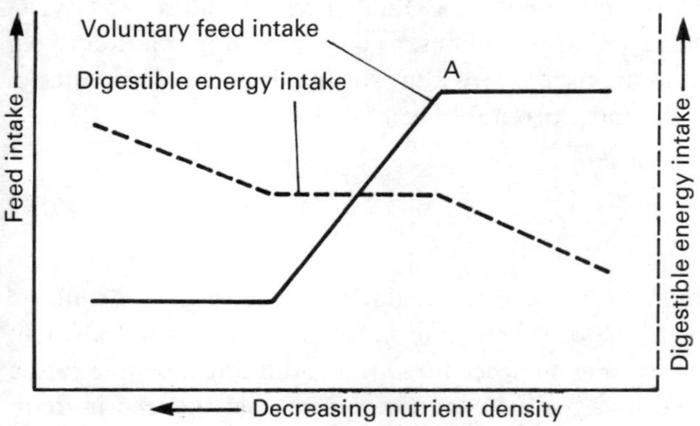

Fig 38 *Relationship between the concentration of DE in the diet and the total daily feed intake (After Cole, Hardy and Lewis, 1971)*

energy, density. Thus a pig will eat more of a diet of lower nutrient density in order to maintain its energy intake, but this will only occur up to a certain limit (see point A, Figure 38). Thereafter, feed intake will not increase any further with dilution of the energy density, and energy intakes will drop (Figure 38).

The art of feeding pigs in the tropics is to maximise use of cheaper feed sources of lower energy density while maintaining adequate energy intakes. The position is also complicated by the fact that high ambient temperatures and irregular water supplies will serve to further reduce the pig's voluntary food intake. Palatability of the food source must also be considered, as a high protein content in a food is of no use to the pig if it cannot be induced to eat it.

Formulation of diets

Objectives

The first step in formulating a ration is to define and decide on the nutritional needs of the pigs. Estimates of nutrient requirements for pigs are available from various sources (see Appendix A). Secondly, the raw materials which are available need to be listed with their nutrient composition. Then the skill is, with due consideration of the factors which have been considered in this chapter, to combine the ingredients into a suitable ration which will provide for the requirements of the pig at least cost. Allowance should always be made for a margin of safety to cover inexact information about ingredients. Moreover, it is very important in a tropical environment to maintain a flexible approach to formulating a ration, according to resources and conditions. Published estimates of requirements should be considered as guidelines, to be modified in relation to pig performance. In economic terms, providing for maximum performance of the pig may not be the most profitable course of action.

Hand formulation

Diets can be formulated by hand calculations, where proportions of ingredients are varied on a trial and error basis until the desired nutrient requirements are obtained. The procedure for formulating a simple ration can best be illustrated by a simple example. Firstly, list the feed ingredients which are available to the producer, noting any constraints which may exist (see Table 5.2). (These details can be obtained from Appendix B).

Table 5.2 Nutrient composition of some typical food ingredients available for a pig diet

	Digestible energy MJ/kg	Crude protein %	Calcium %	Phosphorus %
Maize	13·9	8·9	0·01	0·25
Soyabean meal	12·5	46·0	0·25	0·60
Chickpea meal*	10·8	20·1	0·26	0·32
Bonemeal	–	–	22·0	9·0

*NB Chickpea meal is only available in small quantities.

Then, if a grower ration is required, list the nutrient requirements for growing pigs (Table 5.3).

Table 5.3 The major nutrient requirements in the diet for a growing pig

Digestible energy MJ/kg	13·0
Crude protein %	16·0
Calcium %	0·7
Phosphorus %	0·55

The approximate proportions of ingredients to meet these requirements, based on conventional diets, are known to be:

Energy sources	65–75 %
Protein sources	20–25 %
Calcium/Phosphorus sources	2–3 %
Mineral/Vitamin additives + salt	1·5–2 %

Thus a first attempt to formulate a suitable diet for grower pigs is shown in Table 5.4. Note the amount of nutrient which each ingredient contributes to the final ration is calculated on a percentage basis, e.g. maize is 72 per cent of the ration and contains 13·9 MJ/kg of digestible energy. Therefore, the amount of energy from maize is

$$13{\cdot}9 \times \frac{72}{100} = 10{\cdot}0 \text{ MJ/kg}$$

Table 5.4 A possible ration for grower pigs

	Content (kg/tonne)	Digestible energy MJ/kg	Crude protein %	Calcium %	Phosphorus %
Maize	720	10·0	6·4	0·007	0·18
Soyabean meal	210	2·6	9·6	0·05	0·13
Chickpea meal	30	0·3	0·6	0·007	0·009
Bonemeal	25	–	–	0·55	0·23
Mineral/Vitamins + salt	15	–	–	–	–
Total	1000	12·9	16·6	0·61	0·55

This first attempt at a ration shows that in comparison with the estimated requirements for growing pigs (Table 5.3), the energy is slightly low, the protein content is too high, and the calcium content is also slightly low.

As protein is likely to be the most expensive ingredient in the diet, it is important that the protein content is correct. A useful technique for balancing the protein content is known as Pearson's Square (see Figure 39). The protein level required in the diet (i.e. 16 per cent) is placed in the middle of the square, and the percentage protein content of the ingredient feeds in the two left-hand corners of the square. By subtracting diagonally across the square (16 − 8·9 = 7·1 and 42·8 − 16 = 26·8 in Figure 39) the proportions of ingredients required to give a 16 per cent protein content in the final diet can be read off the right-hand side, i.e. 7·1 soyabean/chickpea to 26.8 maize or 1 part :3·77. By rounding off the figures, this gives the proportions for a practical ration, as in Table 5.5, which provides for the requirements of the growing pig. Suitable mineral/vitamin pre-mixes should be purchased and added to the ration in order to provide for the estimated requirements (see Appendix A).

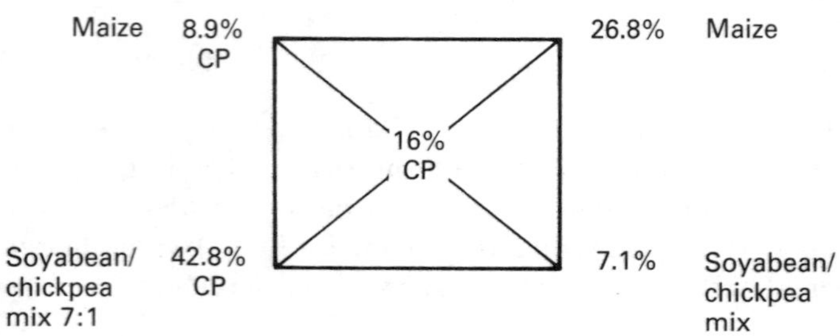

Fig 39 *The Pearson's Square technique for calculating protein levels in diets*

Table 5.5 A corrected ration for growing pigs

	Content (kg/tonne)	Digestible energy MJ/kg	Crude protein %	Calcium %	Phosphorus %
Maize	750	10·4	6·7	0·007	0·19
Soyabean meal	184	2·3	8·6	0·06	0·11
Chickpea meal	26	0·3	0·5	0·007	0·008
Bonemeal	26	–	–	0·572	0·24
Mineral/Vitamins + salt	14	–	–	–	–
Total	1000	13·0	15·8	0·71	0·55

Computer formulation

Ration formulation lends itself to computer programming, and most feed companies and larger operations now formulate on a computer. A database of available ingredients and their cost can be maintained in the computer, and the computer programme can then rapidly solve a whole series of trial and error operations to arrive at the required diet specifications at minimum cost. It is imperative, however, that least-cost ration programming is carried out by a nutritionist, who can check that the end result makes 'nutritional common sense'. The cheapest ration is of no use if the pig finds it unpalatable! (See the companion book on poultry in this series for a more detailed treatment of computer formulation.)

Feed resources

Introduction

The nutrient composition of the feedstuffs likely to be available for pig-feeding in the tropics is given in Appendix B.

Feedstuffs can be classified into energy or protein sources according to whether they are rich in energy or protein. However, energy sources will contain some protein. Energy sources normally form the major part of the diet and are therefore required in larger quantities and will consequently exert a greater influence on overall cost of the ration.

Energy is generally supplied from cereal grains and by-products, root-crops, fats, sugar by-products or various fruits. Protein sources are oilseed meals, fish meals, animal by-products and various plants and nuts. Most of these sources contain some minerals and vitamins, and the remaining requirements need to be met by commercial mineral/vitamin supplements.

Energy sources

The major energy sources are:

Maize

Maize grain is probably the most extensively used energy source, certainly in Africa. It is very palatable and has a high energy value. Lysine and then tryptophan are the first two limiting amino-acids when it is used for pig-feeding. If used at high concentrations on restricted diets, the oil content can cause soft-fat in the carcass.

By-products of maize, after the grain has been processed for human use, namely bran, germ meal, hominy chop, etc., are of higher fibre and lower energy content but can be effectively incorporated in pig diets.

Sorghum (Guinea corn)

As it grows more effectively under drier conditions than maize, it is often used in drier regions. Trials in Zimbabwe indicate it has approximately 95 per cent of the feeding value of maize. Some varieties have a high tannin content which lowers the palatability and protein digestibility of the grain.

Millets

If properly ground, these make an excellent direct substitute for maize, with a tendency to produce firmer fat in the carcass. Due to its susceptibility to infection with the ergot fungus, bulrush or pearl millet should not be fed to breeding sows as the ergot will cause agalactia (lack of milk).

Wheat

Although largely used for human food, wheat has a feeding value equal to maize for pigs if not ground too finely. Where wheat is milled for human use, the by-products such as bran, middlings and weatings are of considerably lower energy content but can be used effectively in pig feeds.

Cassava (Manioc, tapioca)

Production of cassava is widespread throughout Latin America and parts of Africa, mainly on small farms as a staple human food. It has a high energy but very low protein content and this factor must be taken into consideration when cassava is used as a substitute for cereals in pig diets. Large amounts of the meal are exported from South East Asia to Europe.

The roots, which contain approximately 35 per cent dry matter, can be fed fresh, and form a particularly useful source of energy for dry sows. However, care must be taken as some varieties contain cyanogenetic glucosides which can cause prussic acid poisoning.

When the roots are chopped, dried and ground, any free cyanide is destroyed. Research in Zimbabwe has shown that the cassava can replace maize in pig rations with no loss in palatability, but extra protein must be added to the ration. Dried cassava leaves can also be used as an additional source of protein.

Sweet potatoes

These are also widely grown on a small scale. In the fresh state, the tubers contain approximately 32 per cent dry matter and are best used as a foodstuff for sows. When dried and ground, the meal is similar in energy content to maize meal but very low in protein.

Potatoes

Potatoes which are surplus to the requirements of, or unfit for, human consumption can be used for pig-feeding. They contain about 22 per cent dry matter, high levels of energy in the form of starch and protein of very low biological value.

It is important that potatoes are cooked before feeding to pigs. This is because they contain a toxin, solanin, and a trypsin inhibitor which reduces protein digestibility. Both of these are destroyed by cooking.

Yams

Yams are grown extensively for human feed in West Africa and are not generally available for feeding pigs. They contain high levels of starch and in the fresh state they contain both an alkaloid, dioscorine, which is toxic, and tannins which give a bitter taste and reduce digestibility. They therefore should be cooked before feeding to pigs.

Rice

Rice has about 75 per cent the feeding value of maize for pigs and contains a relatively high fibre content. However, in South East Asia and Madagascar the grain is the staple food crop and normally too valuable to use for pigs. Rice by-products, on the other hand, are often the sole source of animal feed in smallholder village production systems. These by-products, known as brans, millings and polishings, consist of the germs and outer layers of the grain. These can represent eight to 15 per cent of the weight of the crop, and their use as pig feed in part explains the importance of pigs in these rice growing areas.

Industrial rice will produce a high quality by-product, whereas traditional technologies do not completely separate the husks and brans resulting in a product of higher fibre content and an excess of silica. 'Coarse' rice brans which include the germ are rich in polyunsaturated fats. This can lead to oily fat in the carcass if fed at a high level to pigs.

Molasses
This is a by-product of the sugar industry and contains highly digestible sugars. It is highly palatable, and can be used to improve the palatability of diets. Molasses can also be a good source of minerals and the B vitamins. However, it has a laxative effect if fed in large quantities and should generally be restricted to under 10 per cent of the diet.

Citrus pulp
This is available as an energy source in citrus-producing areas. It has been assessed to have 47 per cent of the feeding value of maize when used at low levels in the diet. As it can be the cause of diarrhoea, it is not recommended to include it above 10 per cent of the ration.

Fats and oils
In the tropics, both animal fat and vegetable oils are generally far too much in demand by the human population to be used to any extent in pig feeds. Nevertheless, they do form an excellent form of energy and also confer other benefits to the ration in terms of binding the dust, providing the essential fatty acids and giving a mild laxative action. If extraction facilities are not available, whole oilseeds can be used for pig-feeding. Full-fat soyabean, for example, is a good source of protein and energy. Because the raw soyabean contains anti-nutritional factors such as trypsin inhibitors and the toxin, soyin, it has to be heated in order to destroy these factors. In Zimbabwe, small processing plants have been developed for heat-treating whole soyas and render them suitable for feeding to pigs. In smallholder village systems, soyabean can be detoxified by boiling.

Protein sources

Soyabean meal
Extracted soyabean meal is one of the most widely-used protein sources for pigs, and it possesses the highest biological value of the vegetable proteins. The first limiting amino-acid in soyabean meal is methionine, which makes it an ideal combination with maize.

Anti-nutritional factors in the raw soyabean are destroyed by the heat of the oil-extraction process.

Groundnut meal
Groundnut meal is one of the most widely-used sources of protein for pig-feeding in the tropics. Although high in protein content, the biological value is relatively low. It is particularly deficient in lysine and methionine.

Due to the susceptibility of groundnuts to infection with moulds, the meal often contains aflatoxins which seriously depress the growth rate of pigs. In some countries (e.g. Madagascar) village oil mills produce a meal product with a high oil and energy value. This serves as a very convenient source of protein for smallholder systems of pig production.

Cottonseed meal

Cottonseed meal is deficient in lysine and tryptophan and is not an ideal protein supplement for pigs.

Raw cottonseed contains a toxin, gossypol, which can prove fatal. Although this is largely inactivated by the oil extraction process, it is safest not to include the meal above 10 per cent in pig rations. If necessary, gossypol can be neutralised by the addition of 1 per cent iron sulphate to the meal.

Gossypol-free varieties of cotton are presently distributed to farmers in Central and West Africa, and this is likely to increase the use of cottonseed meal in pig diets in the future.

Sunflower meal

Sunflowers are grown fairly extensively by small-scale farmers. The meal is high in fibre (11–13 per cent), and the protein is very deficient in lysine, and is therefore only recommended for inclusion in pig diets at low rates.

Sesame meal

Extracted sesame meal contains 40–45 per cent protein, which is a poor source of lysine but has high levels of both methionine and tryptophan. It is therefore very useful for blending with other protein sources.

Coconut oil meal

Coconut oil meal has about 20 per cent protein content with high fibre. Protein digestibility is often low, and together with a deficiency of lysine, growth will be depressed unless it is only used at relatively low inclusion rates. In some situations, as in the Pacific islands, coconut meal may be the only locally-available by-product containing protein. It may then be used extensively in pig diets.

Palm kernel meal

An increasing hectarage is being devoted to oil-palm cultivation in the tropics, therefore the meal which remains after oil extraction from the oil-palm seeds is likely to become more plentiful. However, because the meal is high in fibre, relatively unpalatable and contains protein of only medium biological value, it should be used only in small amounts.

Safflower meal

Due to the high proportion of hulls in the meal, the high indigestible fibre content severely limits the use of the meal in pig diets. The feeding value is estimated to be about 60 per cent that of soyabean meal.

Beans and peas

There are various species of peas and beans grown throughout the tropics which can be used for pig-feeding, e.g. kidney beans, haricot beans, jackbeans, cowpeas, chickpeas, and pigeon peas.

They all have a similar protein content (18–25 per cent), and are generally deficient in methionine and sometimes tryptophan. Anti-nutritional factors are common in most species and need to be deactivated by cooking.

Rubber tree seeds

In low-input pig production systems in South East Asia, pigs are left to scavenge for rubber-tree seeds when they are shed. Although the seeds may contain toxic cyanic acid when fresh, this is eliminated with age, and the pigs learn to harvest only 'mature' seeds. The seeds contain 17 per cent protein and 18 per cent edible oil.

Fishmeals

These are excellent sources of protein for pigs, with a high biological value and high lysine content. They are also palatable and a good source of calcium and phosphorus. However, their use in the tropics has declined in recent times due to lack of availability and cost.

Meat and meat-and-bone meals

Meals produced from abattoir waste products tend to vary considerably in quality and are not generally in large supply. Unless stabilised by an antioxidant, they also have the problem of the fat going rancid and reducing palatability. Nevertheless, if of good quality, they can make an excellent supplement for cereal-based diets, and are generally included in pig diets at between 5 and 10 per cent. Care must be taken that they are processed properly and do not transmit diseases such as anthrax (see Chapter 7).

Bloodmeals

These are of relatively low palatability, which limits their use in pig diets. However, they can be a valuable supplement as they are particularly rich in lysine, but very often the protein and amino-acids are damaged in the drying process.

Feather meal

Feather meal, a by-product of the poultry industry, has a very high protein content (85 per cent). However, it needs to be hydrolysed, otherwise the keratin protects the protein from digestion. At a low inclusion rate, feather meal can be useful as it has an exceptionally high content of cystine, threonine and arginine.

Milk products

Milk is normally too much in demand for human consumption in the tropics to be used for pig feeds. Both the digestibility and biological value of the protein are very high.

Skimmed milk, which remains after the removal of butter fat from whole milk, can be fed in fluid form or dried to give a powder containing 33 per cent protein. It is particularly useful for inclusion in creep rations for baby pigs to improve the palatability and digestibility.

Whey, which is a by-product formed in the cheese-making process, contains only about 5 per cent dry matter, but can be dried to yield a powder of 15 per cent protein content.

If milk products are fed in the fluid form, they should be fed either consistently fresh or consistently soured.

Some other bulky foods

Bananas

Ripe bananas are very palatable to the pig, and studies have shown that the growing pig will consume up to 8 kg per day. They only contain 1 per cent protein, but can provide a useful supplementary source of energy in the form of sugars.

Brewing by-products

These consist of either wet or dried brewer's grains from barley, sorghum or millets. In the process of brewing, the soluble carbohydrates are utilised to produce alcohol so that the residue is richer in protein and fibre than the original grain.

Although, because of the high fibre content (15 per cent), low protein and great variability, they are of limited value for pigs, they tend to be quite widely used in small-scale systems of pig production, especially in Africa. Residues from traditional beer brewing systems are also often available to pigs on scavenging systems of production.

Water hyacinth

Water hyacinth, which is a troublesome weed of waterways throughout large parts of the tropics, is used as a feed for pigs in South East Asia. The plants are chopped, sometimes mixed with some other vegetable waste, and boiled slowly for a few hours until they turn into a paste.

This is supplemented with oil-cake meal, rice bran and sometimes maize or salt. A common formula is:

40 kg water hyacinth
15 kg rice bran
2·5 kg fish meal
5 kg coconut meal

Leaf protein

Although the high levels of fibre will limit the amount that can be included in pig diets, various green leaves such as lucerne, other legumes, cassava, sweet potato vines and grass can be dried and ground or fed fresh to pigs. In general, the leaves should be as young as possible to minimise the effects of fibre.

Swill

Swill or kitchen waste from institutions, hotels, restaurants, etc. can form a significant source of food for pigs. It tends to be used when pigs are reared in close proximity to a town or city.

The nutrient content of the swill can be very variable according to source. Also, it can transmit infectious diseases (e.g. Swine fever and foot and mouth disease), and it is recommended that all swill is boiled before feeding to pigs. Swill often contains broken glass, fishbones and pieces of metal which can be harmful to pigs.

The best system of feeding swill is to use the 'Lehmann system', which was developed for bulky foods. On this system, the pigs receive a set small amount of balanced concentrate each day and then have *ad lib* access to the swill and they can eat as much as they like.

Pumpkins and melons

Pumpkins and melons grow well in the tropics, and are commonly grown in maize and other crop lands. They are often fed to pigs on small-scale systems. The nutritive value is very low, and they contain virtually no protein, but they can provide some energy and succulence, particularly on scavenger systems of production.

6 Housing

Principles

Pig productivity and comfort

The most productive pigs are likely to be those contained in a thermally neutral environment; that is, when the environmental temperature around the pig is consistently between the pig's lower (LCT) and upper critical temperature (UCT) (see Chapter 2). The pig's metabolic heat production is then at a minimum, and it is neither using feed energy to keep warm, nor reducing feed intake to keep cool.

Other considerations for pig comfort and well-being in addition to temperature are:

- protection from other climatic extremes such as direct sun, wind and rain;
- provision of dry conditions which are hygienic and do not predispose the pig to disease;
- allowing, as far as possible, for inherent behaviour patterns of the pig, and minimising the effect of social dominance;
- provision of accessible food and clean water;
- providing conditions so that good stockmanship can be practised;
- effective disposal of effluent.

General considerations of design

Any buildings, whether simple or complex, cost money, both to build and to maintain. This makes it imperative that careful thought is given to the right design, so that the investment is justified by improved productivity. Moreover, considerations that affect design of houses in the tropics can be very different from those in more temperate parts of the world.

Under tropical conditions, the paramount consideration is generally to ameliorate the effects of excess heat. At the same time, it is important to minimise temperature variations, keeping as close as possible to the pig's zone of thermal neutrality. This often involves keeping pigs cool by day and warm by night, particularly at higher altitudes where diurnal fluctuations in temperature can be as high as 25 Centigrade degrees. In practical terms, therefore, buildings need to be designed to provide for this, while at the same time allowing for the other prerequisites for the well-being of the pig.

Siting

Although the choice of a site is often limited, siting and construction to take advantage of the natural prevailing winds can increase ventilation and air flow. By facing buildings north–south, excessive penetration of sunlight into pens can be avoided, preventing problems of sunburn and heat stress. At the same time, the site should preferably not be too exposed to cold.

Ideally, pig buildings should be on a slight slope to facilitate drainage and disposal of effluent. The slope also makes it easier to design a 'pig-flow', with the farrowing accommodation at the top of the slope and the fattening pens at the bottom. This is likely to improve management and prevent cross-infection from effluent between categories of pigs, an important concept in preventing disease in the most susceptible groups, such as baby pigs.

Other considerations are proximity to a good water supply, ensuring that effluent cannot contaminate water supplies from boreholes or wells, and access to a good road for more intensive piggeries to allow for transport of food in and pigs out. Finally, allowance should be made for adequate space between buildings and for possible expansion in the future.

Floors

Even in low-cost buildings, hard concrete or similar floors are recommended. This prevents the pigs 'rooting' and digging up the ground and most importantly allows for easy cleaning and the prevention of the build-up of disease and parasite problems.

The floor should provide insulation against both the cold and damp. Many insulating materials can be used, including those which provide an underfloor airspace such as old bottles, air bricks, piping, etc., but stone or gravel is most common. A typical suitable piggery floor is shown in Figure 40.

The nature of the floor surface is important. If the surface is too smooth and slippery, this will result in injuries. If it is too rough or abrasive, this can cause damage to feet and udders.

Roofing

Roofs need to provide shade and protection. They can be extremely simple and made from local materials such as grass, reeds, leaves, etc.

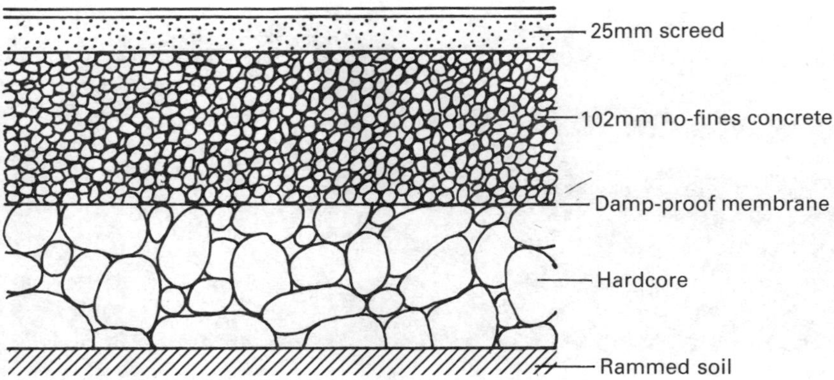

Fig 40 *A suitable piggery floor*

Fig 41 *Illustration of simple low-cost housing made from local materials*

(Figure 41), or from manufactured roofing material such as tin or asbestos. A thatched grass roof, for example, makes an excellent roof under hot conditions as it is well-insulated. Unfortunately, however, if not protected, it rapidly becomes a breeding ground for rats and other pests.

In hot climates, asbestos or tin roofs should be insulated with a layer of grass, insulating board or equivalent, or be sufficiently high that there is room for plenty of air movement. On an apex roof, there should be a raised ridge (see Figure 42) or other arrangement to allow for the escape of hot air and air movement.

Fig 42 *A raised roof ridge allows for hot air to escape*

Fig 43 *Flaps allow for maximum ventilation and for containing warmth when ambient temperatures drop*

Walls

External walls should be of adequate construction to contain the pigs and provide protection. They can be built of several materials, ranging from mud and/or sticks (see Figure 41) through to bricks and cement.

In enclosed buildings, it is very important to provide suitable flaps (see Figure 43), windows and doors which can be opened to allow maximum ventilation during the hot periods, and closed to contain warmth at colder times.

Space

Recommended space requirements for various categories of pigs are shown in Table 6.1. Overcrowding is a common cause of depressed performance and productivity in pigs.

Table 6.1 Recommended space requirements for pigs

	Area
Boars	9 m^2
Dry sows	
Stalls	2 m long × 0·64 m wide
Cubicles	As for stalls plus similar exercise area
Yards	3–4 m^2 per sow
Farrowing accommodation	
Pen, including crate	6·2 m^2
Fattening/follow-on (including creep area)	10 m^2
Weaners	
Cages (per pig)	0·2 m^2 lying area + 0·2 m^2 slatted area
Yards (per pig)	0·7–0·9 m^2
Porkers	
(Pen, including dunging area)	0·73 m^2 per pig
Baconers	
(Pen, including dunging area)	0·93 m^2 per pig
Heavy pigs	1·10 m^2 per pig
Trough space (per pig)	
Fatteners	0·2–0·3 m^2
Maiden gilts, sows	0·35 m^2

Boar housing

Protection of boars against temperature extremes is very important, as high temperatures can rapidly lead to lower fertility of semen and reduced libido.

Strongly constructed boar accommodation is essential (see Figure 44), particularly to prevent boars getting out and fighting. Wire reinforcement can be used for walls, and bolts and hinges for doors must be robust.

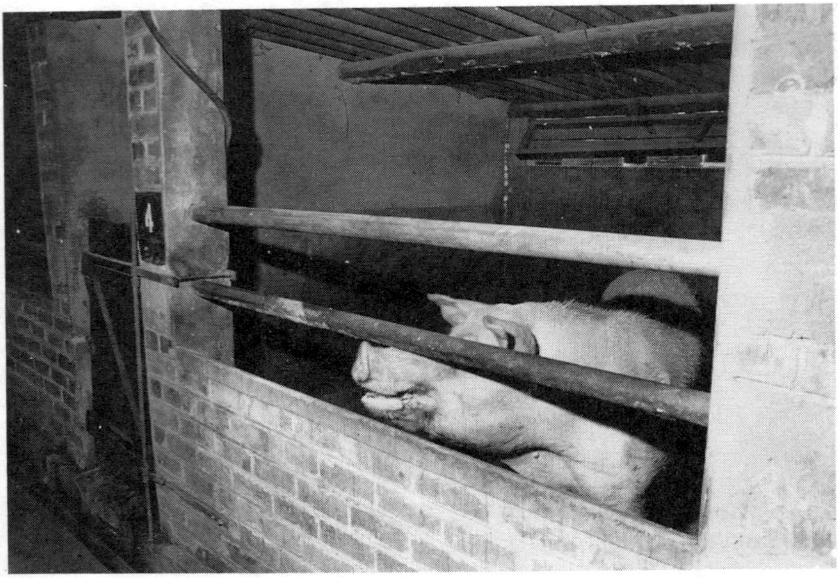

Fig 44 *Strongly constructed boar housing is absolutely essential*

Sows should not be served by the boar in the boar pen. A simple service area should be constructed outside (Figure 45). This has the advantage of being cooler (early morning and evening), providing a good foothold to the pigs on the grass or soil and it obviates any right-angle corners which can interfere with service procedure.

Dry sow housing

Yards

Yards are the simplest form of dry sow housing, and consist of a yard or large pen shared by several sows. In order to avoid bullying and fighting, sow groups should not be too large. Ample shaded lying area and trough space must be provided. Ideally, sows should have access to individual feeders so that feed intake for individual sows can be controlled.

This system is widely used by small-scale and less intensive producers. It is the best system for gilts as it allows them to interact and stimulate oestrous behaviour.

Tethers

The system of tethering sows on long leads, which is commonly seen throughout the tropics, can be used within yards or paddocks as a means of restricting sows to a given area.

Fig 45 *A simply constructed service area, providing good conditions for service*

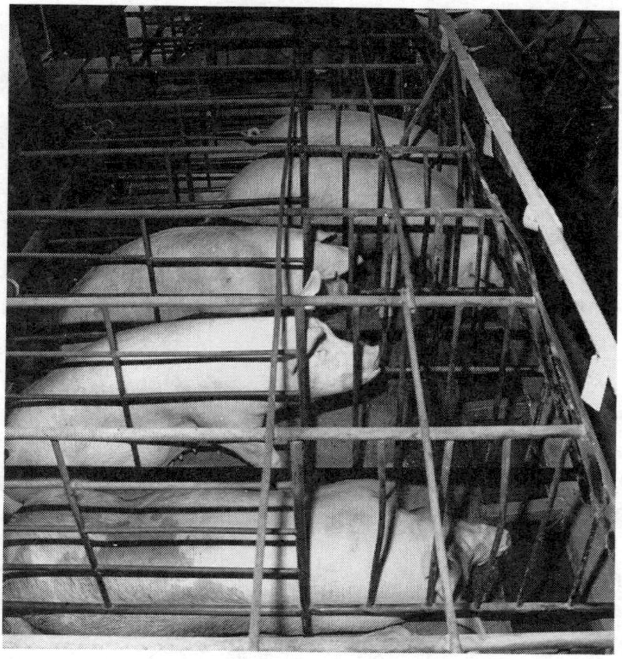

Fig 46 *Individual sow stalls for pregnant sows*

Sow stalls

In sow stalls, each sow is individually penned and fed in a small pen which restricts movement (Figure 46). The main advantages of individual stalls are that bullying and fighting are eliminated, cleaning-out, feeding and management are facilitated, sows can be fed according to individual requirements and the sow does not use up energy in walking. The major disadvantage is that it can lead to lameness and lack of muscle tone in sows.

Care must be taken in design that sows are well-protected from the weather on all sides of the building, as they cannot huddle together to conserve body warmth in cold conditions.

Cubicle housing

In this type of accommodation, sows have a communal dunging and exercise area and each has an individual stall for lying and feeding (Figure 47). The dunging passage can also be used very effectively for the boar to make his daily rounds for heat detection, and the design normally incorporates boars' quarters. Cubicles tend to combine the advantages of sow stalls (individual feeding and attention) and yards (exercise and comfort).

Electronic sow feeders

The advent of computer-controlled electronic sow-feeding systems in the developed countries has provided the potential for incorporating the advantage of yards with individual sow-feeding. Sows housed communally can gain access to individual feeders by means of transponders installed around their necks or in their ears. This level of sophistication is unlikely to be in general use in developing areas of the world for some time to come.

Farrowing and rearing accommodation

This is probably the most important building in the piggery, providing for the birth of piglets and getting them off to a good start in life. The need for specialist design in farrowing pens has increased as sows have been bred with larger body sizes, bigger litters and reduced mothering ability. Thus, while the small indigenous sow will make a nest and farrow and rear her piglets very efficiently without any help, for the average exotic sow, piglet survival will decrease unless certain provisions are made. The essential features of farrowing house design are:

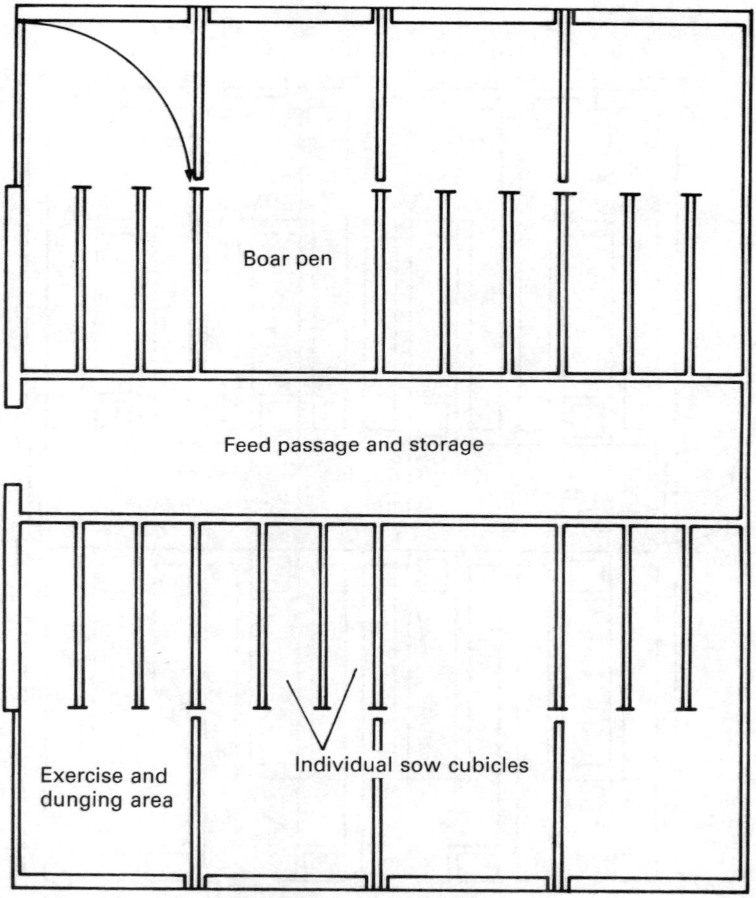

Fig 47 *Plan of cubicle house*

- protection of the piglets from being overlaid by the sow;
- provision of higher temperatures for the piglets than for the sow. If the whole house is the ideal temperature for the piglets, this will reduce sow appetite;
- easy access for cleaning so that high levels of hygiene can be maintained.

Farrowing crates

These are specialist crates, which are generally incorporated into specialist farrowing houses (see plan in Figure 48). The sow is restrained in the crate, while piglets have access to a small surrounding pen which includes

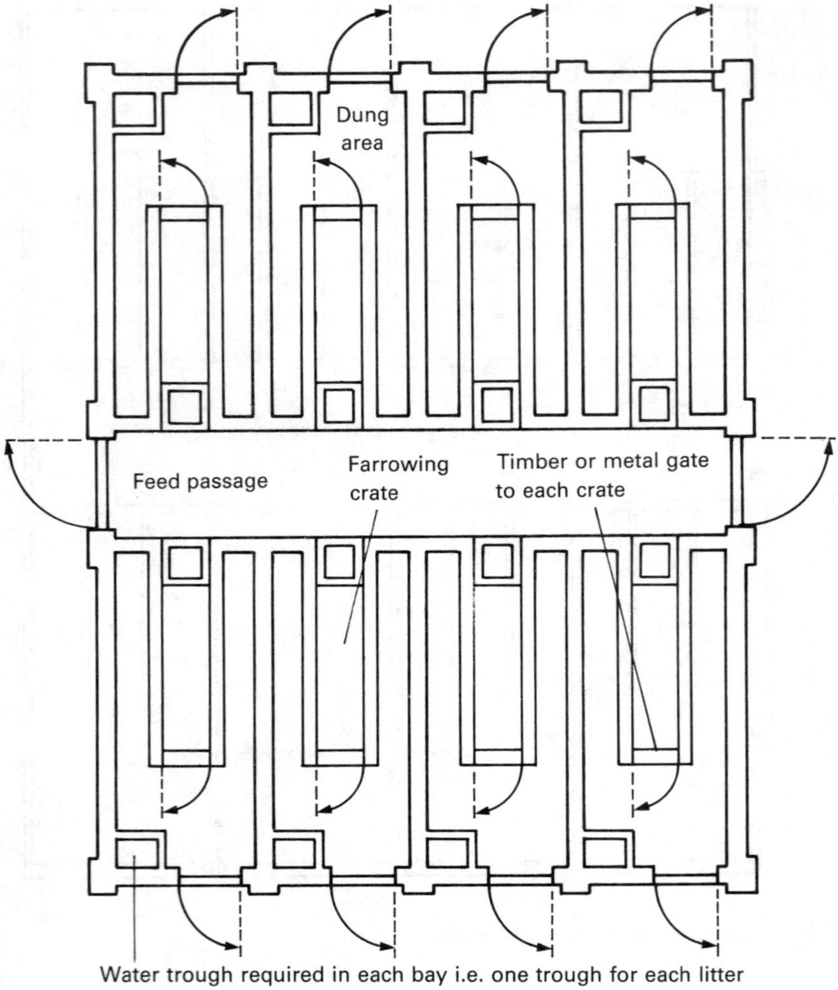

Fig 48 *Plan of farrowing house*

a creep box, creep feeding facilities and their own water supply. The design and dimensions of the crate are important to ensure both sow comfort and farrowing efficiency (Figures 49 and 50).

The major advantages which accrue from the use of a crate are:

- restraining the sow from lying down quickly and crushing piglets;
- provision of creep boxes close to the sow, which is very important for the first few days after birth;
- allowing for the stockman to handle the sow and her piglets without risk of injury;
- free access to creep feed and water by the piglets.

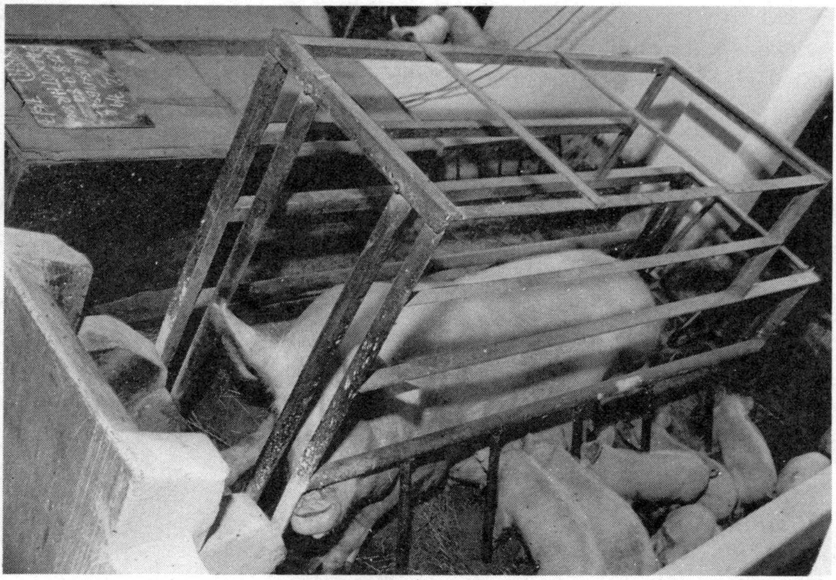

Fig 49 *A specialised farrowing pen containing a farrowing crate and creep box*

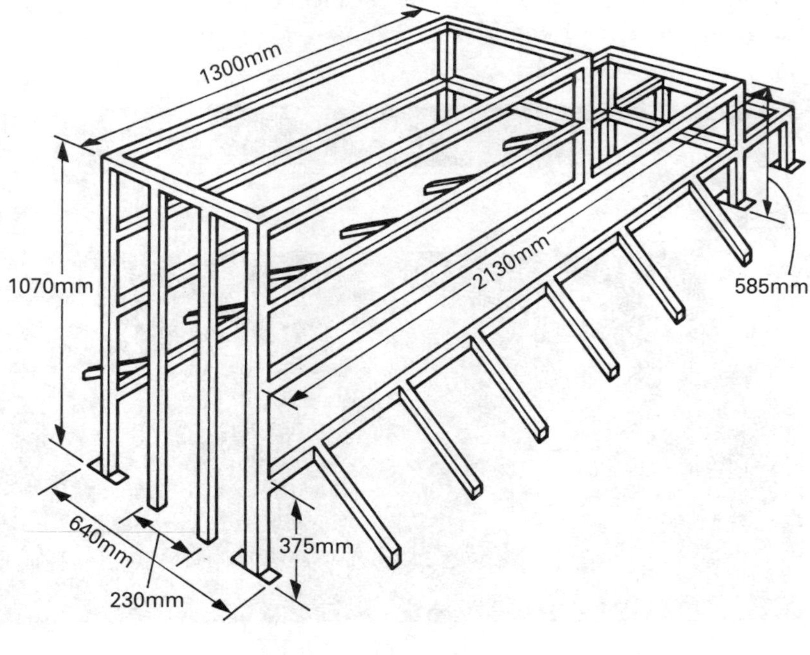

Fig 50 *Farrowing crate*

Even under tropical conditions, a separate creep area for the piglets which is warmer than the ambient temperature is generally an advantage, especially at night. This is because the optimum environmental temperature for the sow is between 16 and 18°C, whereas that of the newborn pig is 33–35°C. A simple, enclosed creep box (Figure 49) is perfectly adequate for the piglets to creep into and generate their own warmth by huddling together. If electricity is available, or a paraffin source, a light or simple heater can be provided in the creep box. This not only provides extra warmth, but attracts the piglets into the box and away from the danger of being crushed by the sow.

If separate arrangements are not made for the piglets and the whole farrowing room is warm, this reduces sow productivity. American work has shown that for every °C rise in temperature from 25°–30°C, daily feed-intake by the sow declines by 400 grams.

Multi-purpose pig pens

These are liable to be more appropriate to the tropics and the developing world, as they are cheaper and more flexible. Removable structures, such as creep barriers and farrowing rails, provide protection for the piglets and make the pen suitable for farrowing. At weaning, these are removed, leaving a fattening pen in which the weaners can be fattened through to slaughter (Figures 51 and 52).

Fig 51 *A multi-purpose pen*

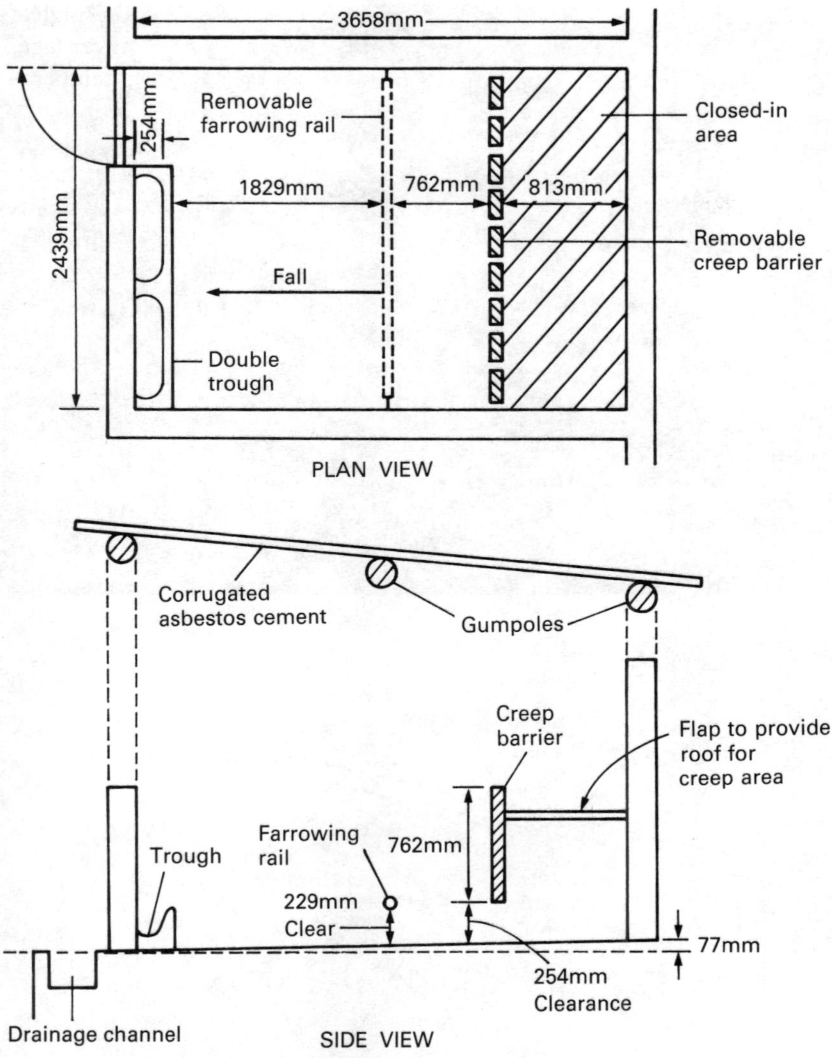

Fig 52 *A multi-purpose pen. Farrowing structures are removable so that the pen can be used for fatteners or other stock*

Follow-on pens

If specialised farrowing houses are used, sows can normally be removed into cheaper housing after 10–14 days when all danger of sow-related piglet deaths is past. Piglets then require a separate creep area, similar to that provided by a multi-purpose pen. This has the advantage of allowing the sow to exercise and move around freely.

Fig 53 *The front of a weaner cage*

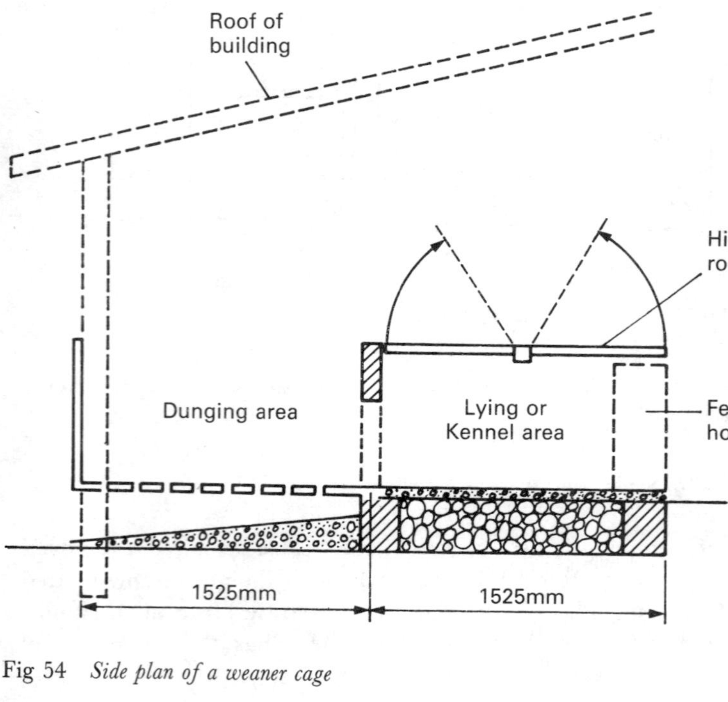

Fig 54 *Side plan of a weaner cage*

Weaner cages

The combined trauma of weaning from the sow and change in diet makes the young pig very susceptible to disease, particularly digestive diseases. This can result in fairly heavy mortality of weaners. The weaner cage was originally designed in Europe with the ideal of providing conditions for the weaner which would help overcome these problems. Weaner cages have since been adapted for hotter climates and basically consist of a covered solid-floor sleeping and eating area and a dunging area floored by either wooden slats or metal mesh (Figures 53 and 54). During cold periods, pigs can huddle and generate enough heat for their comfort inside the covered kennel section. Ventilation is provided by the centrally-hinged roof of the kennel (Figure 54). In hot weather, pigs can keep cool by lying out on the meshed floors, and are protected from the sun by an umbrella roof over all the cages. As dung and urine falls through the wire mesh or slats, this can be cleaned from below and there is therefore no need for stockmen to enter the cage with contaminated boots, brooms or shovels.

Pigs normally remain in the cages for three to four weeks before being transferred to fattening accommodation. The feed hoppers (see Figure 54) can be moved to allow additional space as the weaners grow.

Weaner pools

The traditional system of housing weaners is to take litters of similar age and move them into large pens holding up to 50 weaners. After 3–4 weeks, pigs are batched into groups of equal sizes for transfer into growing/fattening pens.

Ample watering and feed-trough space must be provided, and some form of bedding is preferable. A sheltered kennel area, which can be insulated by a ceiling of hay bales or wood, can provide for added warmth.

Growing/fattening accommodation

The basic needs for good fattening pens are relatively simple, namely, a dry lying area and a demarcated dunging area (see Figure 55). The building should provide shade, some protection and adequate ventilation. Solid walls are not required between pens, as they will decrease ventilation and air flow within the building. Pens designed to hold 8–10 pigs through to slaughter are the ideal size (Figure 55).

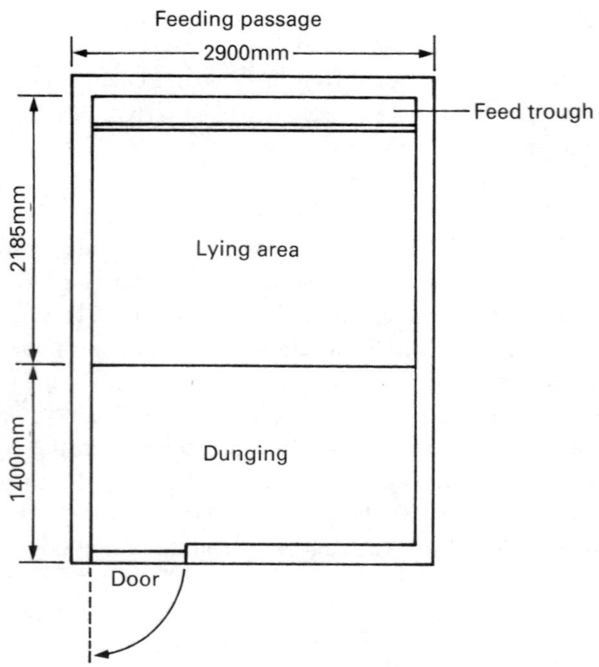

Fig 55 *Floor-plan of a growing/fattening pen designed to hold 8–10 fatteners*

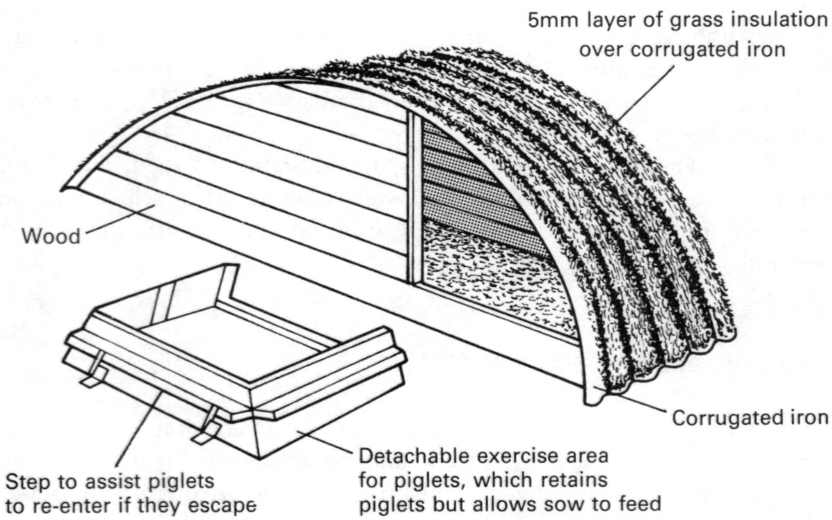

Fig 56 *A farrowing ark for use under extensive systems*

Fig 57 *A well-constructed wallow*

Extensive systems

Extensive systems (see Chapter 3) are particularly appropriate for sows. Sows are run in paddocks and have access to arks or huts in which to farrow. In trials in Zimbabwe, sows were allowed a choice of different designs of arks at farrowing time, and it was found that they preferred a design similar to those found in the UK (Figure 56). The major difference is that under tropical conditions, the roof should be insulated with a 5 cm layer of grass or similar material. Arks can be constructed from cheaper materials, but it is difficult to make them sufficiently robust to avoid destruction by the sow. Ample deep shade and wallows (Figure 57) should also be provided for sows run under this system. As mentioned previously, tethers can be used as a means of restricting sows within a paddock. They can then be rotated around a given area of pasture or other forage.

7 Health

Disease prevention

Once disease affects a pig herd the impact on the economics of pig production in terms of the cost of control and decreased productivity can be enormous. The first priority must therefore always be to try to prevent the occurrence of disease. Thus, many of the management procedures considered in Chapter 8 are aimed at disease prevention or at mitigating the effects of those diseases that cannot be prevented. With skilled management, combined with well-designed housing and sound nutrition, an overall strategy to minimise the possibility of disease attack can be formulated.

At the same time a basic knowledge of the main diseases which may affect a pig herd is necessary so that a producer can diagnose the condition and implement control measures as quickly as possible. This is of particular significance under tropical conditions where the regular services of a veterinarian are often not available. The major disease problems are parasites, infectious diseases and a few non-specific diseases. Nutritional deficiency conditions have been dealt with in Chapter 5.

Parasites

Parasites are defined as organisms which live on and obtain food from the body of another, known as the host. They may live on the exterior of the pig, when they are known as external parasites, or within the internal tissues and organs when they are known as internal parasites. Parasites will seldom result in the death of the host except in the case of massive infestations or if the host is also stressed in other ways.

External parasites

These mainly cause irritation to the skin surface, often leading to wounds and an increased susceptibility to other infections. The most common external parasites are mange-mites, ticks, lice, fleas and flies.

Mange-mites

Mites, which are scarcely visible to the naked eye, spend their entire life-cycle under the skin of the pig, but they can survive off the host for as long as eight days. The most common species is *Sarcoptes scabiei* which causes sarcoptic mange.

First signs of infection are a crusty, dry-looking skin around the eyes, ears and snout. The mites then spread and multiply over the body, and their burrowing causes the skin to become inflamed and swollen. The pig will be seen to be constantly rubbing itself and performance is depressed.

Control is best effected by regular treatment, either dipping or spraying with an anti-mange medication, including spraying of pens. Chronically infected animals should be culled. There are also some recent systemic drugs on the market which are very effective against the mite.

Ticks

Ticks are only a problem in scavenging or more extensive systems of pig production. There are a number of different species which suck blood and can transmit serious diseases (e.g. Babesiosis or redwater). They generally require more than one host to complete their life cycle. Ticks are easily controlled by spraying or dipping with suitable acaricides.

Lice and fleas

Both lice and fleas can become a problem in dirty and unhygienic conditions, as they live on the skin surface, suck blood and cause irritation. Spraying of the pigs and pig quarters with suitable insecticides are effective ways of controlling the pests. In the case of lice, particular attention should be paid to the ears.

Flies

Flies have a major nuisance-value around pigs as they cause annoyance, can bite, and carry infectious diseases. They are always attracted to any fresh abrasion or wound on the animal.

Control measures should involve spraying of insecticides on suitable fly-breeding areas, e.g. manure heaps, refuse areas and ponds, pig-buildings and the pigs. Baits which attract the flies and are poisonous to them but not the pigs can also be effective.

Internal parasites

Roundworms

These are a particular hazard when pigs are free-ranging or not kept on concrete floors. The large roundworm (*Ascaris lumbricoides*) is very common and can cause a lot of damage in pig herds. Adults live in the small intestine and can grow up to 300 mm long and 6 mm thick (Figure

58). The female is capable of laying thousands of eggs per day, which pass out in the dung and become infective, if ingested by other pigs, after 21 days. These eggs are extremely resistant and can remain infective for many years. As part of the life-cycle (Figure 59), eggs hatch out in the pig after ingestion and the larvae migrate through the liver and lung. Irritation in the lungs causes coughing and ill-thrift, particularly in younger pigs. Damage is also done to the liver which renders it liable for condemnation at slaughter ('milk-spot liver'). Moreover, if infection is heavy the adult worms can partly obstruct the small intestine, causing weakness and loss of weight by the pigs.

Another harmful worm is the whip-worm (*Trichuris suis*). This worm, which is about 35 mm long when adult, lives in the large intestine and causes considerable damage to the gut wall resulting in diarrhoea and weight loss. The nodular worm (*Oesophagostomum spp.*) also lives in the large intestine. It burrows into the intestinal wall forming nodules, and can cause diarrhoea (sometimes bloody) and anaemia. The kidney worm (*Stephanurus dentatus*) lives in the kidney and eggs are excreted via the urine. When ingested, larvae migrate through the liver to the kidney and tissue damage results. This kidney worm is a major handicap for free ranging pig farming systems in Madagascar and the Pacific Islands, and is often the main reason why pigs are penned. Lungworms (*Metastrongylus spp.*) can also be a problem in free-range pigs; infection occurs when pigs eat earthworms which are the intermediate hosts. Lungworms cause irritation and coughing and predispose the lung to secondary pneumonia.

Contaminated feed and water are the usual source of infection with internal parasites. Control can be effected by breaking the life-cycle, which means regularly moving range pigs on to fresh ground and frequent cleaning and removal of faeces in housed pigs. At the same time, unless there is good evidence that there is no worm infection in the herd, breeding pigs should be routinely dosed with broad spectrum anthelminthics and young stock dosed soon after weaning.

Tapeworms

The common tapeworm is *Taenia solium*. The pig is its intermediate host and the adult worm lives in man. Pigs become infected by picking up eggs from human faeces and the larvae then encyst in the pig's muscle, particularly in the region of the heart and tongue.

If the pig meat is then eaten by man, the larvae hatch out and the cycle is completed. As a consequence, carcasses which are affected (measly pork) are condemned at slaughter. By preventing pigs having access to human faeces, the parasites can be eliminated. In some countries live pigs are checked at the market place by trusted experts for the presence of tapework cysts in the tongue. The result of the examination influences the price paid to the producer.

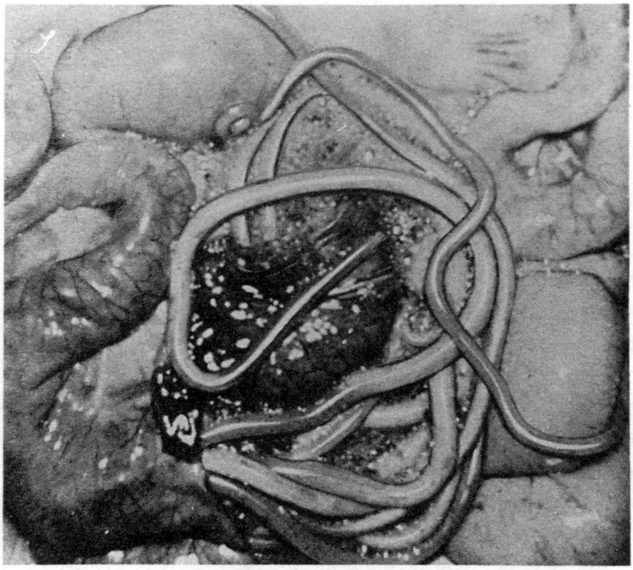

Fig 58 *Adult Ascarid worms in the small intestine*

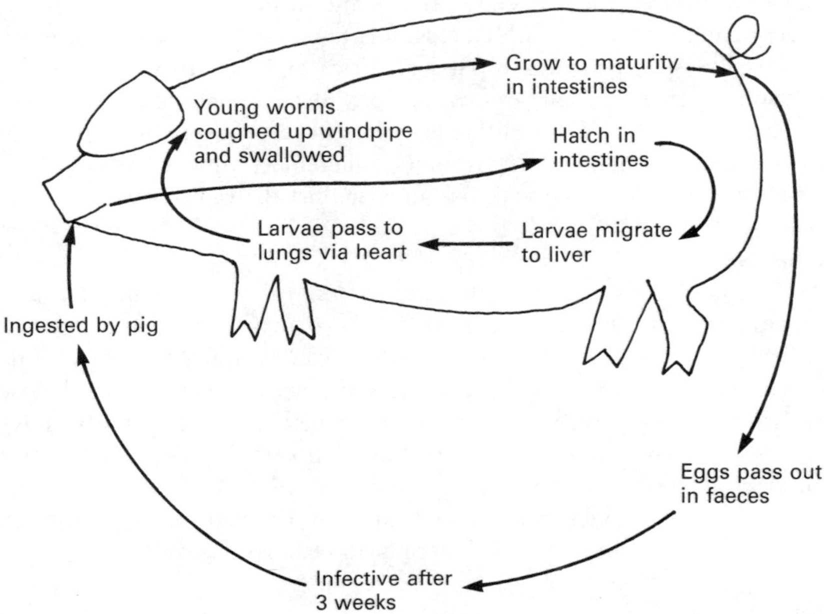

Fig 59 *Life-cycle of the Ascarid worm*

Infectious diseases

The following diseases are notifiable in most countries.

African swine fever

This is a highly contagious virus disease which in the acute form can cause 100 per cent mortality. Typical symptoms are loss of appetite, pigs huddling together, small purplish blotches on the skin, incoordination and laboured breathing.

In Africa, both bush pigs and warthogs are carriers of the virus but are immune to the disease, and it is therefore very important to prevent direct contact between domestic pigs and wild species. This contact can be prevented by double penning and the control of animal movements. Moreover a soft tick (*Ornithodoros moubata*) which infests the warthog, is a biological carrier. Otherwise infection occurs by contact with other sick pigs, or through contaminated food or water. There is no effective vaccine or treatment and infected pigs should be isolated from healthy ones. Although the disease originated in East Africa, it is gradually spreading west through Africa.

Foot-and-mouth disease

Regarded as the most contagious of all known viral diseases, infection causes blisters on the feet, snout, udder and in the mouth and throat. It is very painful to the pig, which cannot eat and often has to be destroyed.

The disease is endemic in parts of Africa and the virus is carried by the buffalo. Infection can occur by feeding infected bones or cooked meat.

There is no cure. If an outbreak occurs in an adjacent area, pigs can be vaccinated, but as there are many different strains of the virus it is important to ensure that vaccination is against the right virus.

Anthrax

This is an acute and frequently fatal bacterial disease which often also causes mortality in humans. There are two main types of symptoms, either a swelling in the back region which causes difficulty in breathing or sudden death with blood oozing from the body orifices. If the disease is suspected, the carcass should not be opened as this releases infective spores. Infected carcasses should be buried at sufficient depth to prevent any transmission of the spores.

Pigs are infected by contact with anthrax carcasses or by spores in contaminated food or pasture. There is an effective vaccine.

Rabies

Pigs with rabies show intense excitement, bite other pigs, salivate, become uncoordinated and die within 2–3 days. This disease is also fatal to humans. Infected animals should be shot through the heart, as brain damage can result in release of the virus.

Pigs are infected by contact with the saliva of infected animals, particularly dogs and jackals. Such contact should be prevented.

Trypanosomiasis

Although trypanosomiasis is mainly restricted to the tsetse-fly areas of Africa, pigs are nevertheless susceptible to the disease. It is caused by a protozoan, *Trypanosoma vivax* or *congolense*, which is transmitted by the tsetse fly. Infection will cause debilitation, anaemia and ultimately death. In some cases, the acute form of the disease will cause sporadic sudden death without symptoms. There are, however, effective drugs against the disease but these tend to be costly. Another form of trypanosomiasis occurs in Asia and Tropical America and appears to be on the increase. It is caused by *Trypanosoma evansi*, and is transmitted by species of *Tabanus*, a biting fly.

Other infectious diseases

Brucellosis

This disease, which is caused by a bacterium, is also known as contagious abortion. Brucellosis can result in temporary or permanent sterility in females. Abortion is the most common symptom and can occur at any stage of gestation, depending upon the time of exposure to infection with the bacterium. In boars, testicles may become inflamed and permanent sterility may result.

The disease is transmitted at mating or by contaminated food or water. There is no treatment and infected animals should be culled, particularly as brucellosis is transmissible to humans, and the risks of transmission are relatively high under some traditional systems of pig management. Brucellosis appears to be widespread in pig herds in South East Asia and the Pacific Islands.

Coccidiosis

This is caused by organisms known as coccidia, of which there are 13 known infective species in swine throughout the world. They cause damage to the intestinal wall, and are believed to be an increasing cause of diarrhoea in piglets, particularly in confined housing. Piglets show a grey-green diarrhoea, lose weight and rapidly become dehydrated.

Coccidiosis is spread by contaminated faeces and thus good management and regular cleaning of buildings will prevent the disease. Drugs, known as coccidiostats, are available for prophylaxis and treatment.

Cystitis/nephritis

Cystitis is due to bacterial infection of the bladder and subsequent kidney infection and is particularly prevalent in sows in confined housing. The disease can be recognised by the presence of pus, and later blood, in the urine. Generally species of *Streptococcus* and *Corynebacterium* are involved, and although large doses of antibiotics can sometimes be beneficial, the disease is usually fatal, due to the difficulty of getting the antibiotic to the site of infection.

Antibiotic treatment of the penile sheaths of boars and increasing exercise levels for sows by avoiding close confinement can help prevent the condition.

Enteric colibacillosis

Diarrhoea caused by the bacterium *Escherichia coli* is the most common cause of death in baby pigs. *E. coli* are normal inhabitants of the intestinal tract, particularly the large intestine. Various stress conditions cause these bacteria to multiply rapidly in the small intestine. They produce toxins which stimulate a massive fluid loss into the small intestine, leading to scours and dehydration.

E. coli scours is primarily a problem in baby pigs up to 10 days of age. The sows remain healthy, the piglets will continue to suckle, they develop a severe watery diarrhoea and mortality can be heavy. The infection will spread slowly among newly-born piglets.

The main preventative measure that can be used against *E. coli* scours is good management, i.e. not allowing for stressful conditions in the piglets which will trigger off scouring. Antibiotics given orally can be effective if given immediately symptoms are seen. Vaccines are available which when given to sows and gilts help ensure that new-born piglets receive additional antibodies via the colostrum.

Gut oedema is another condition caused by *E. coli.* It affects pigs after weaning, normally between eight and 25 kg liveweight. The condition is most prevalent in the fastest-growing pigs and the typical symptom is oedema in the region of the stomach and intestines. Pigs begin to stagger and will often die after exhibiting convulsions. The proportion of animals which are affected will vary considerably. It is generally believed that stress is the major cause of gut oedema, particularly overeating of high-protein diets. The disease can be contained by reducing feed-intake, increasing the fibre-level and lowering the protein-content of the ration.

Enzootic pneumonia

This condition, which is caused by mycoplasmas, is widespread in parts of the tropics. Mycoplasmas cannot live for long outside the body and close contact of infected and healthy pigs is the most common form of transmission.

The clinical signs are a raised body temperature and an increased respiration rate accompanied by bouts of coughing. Enzootic pneumonia is a complex condition in which various environmental conditions and stress are involved. The mycoplasmas are resistant to antibiotics and the best policy is to try to keep a herd free from the disease. Antibiotic treatment is useful to prevent the occurrence of secondary infections which can often occur.

Erysipelas

The causative agent of erysipelas is a bacterium which normally lives in the soil. There are three distinct stages of the disease: acute, sub-acute and chronic. In the acute form sudden death is common. Sick pigs show marked constipation, a high temperature (41–42°C) and a reddish/purplish discoloration of the ears, abdomen and legs. The discoloration often shows up in somewhat rectangular raised patches of skin. If treatment is not timely, the sub-acute stage is less serious. Of more importance is the chronic stage which leads to a chronic arthritis, swollen joints and stiffness, and heart damage. Affected joints will be condemned at slaughter.

Once the disease is diagnosed, treatment as soon as possible with antibiotics will cure it. Excellent vaccines are available, and a routine vaccination programme is recommended to prevent infection. The disease is transmitted by animal contact or is picked up from the soil. Because erisipelas, like brucellosis, can be transmitted to humans, care must be taken by people who are in contact with infected animals.

Greasy pig disease (exudative epidermitis)

The disease is caused by an infection of the skin with a skin bacterium, *Staphylococcus hyiens*, and affects pigs from one to seven weeks of age. There is scale-formation on the skin, which give rise to crusts, and the skin becomes greasy with matted hair. There is no fever and pigs do not scratch. In the early stages, the condition can be confused with zinc deficiency (see Chapter 5).

There are effective medications for dipping infected pigs, combined with broad spectrum antibiotic injections.

Mastitis – metritis – agalactia (MMA syndrome)

This syndrome is a common problem which results from bacterial infection in the newly-farrowed sow and may occur as a single entity, or any of the three combined. In sows infected with mastitis, the mammary glands become swollen and inflamed and this may or may not be associated with an infected uterus (metritis). In either case there is invariably a lack of milk flow (agalactia) and piglets will become thin and begin to die on the second or third day due to starvation.

A number of stress factors including inadequate diet, high ambient temperature, nervousness, other disease and retained placenta may be instrumental in predisposing the sow to MMA syndrome. Everything possible must therefore be done to avoid stress in the farrowing quarters. For sows with MMA syndrome, the first essential is to get the milk flowing again and this can be achieved by injections of the hormone oxytocin. Antibiotics can be used to counteract the metritis and mastitis. Adequate exercise and a plentiful water supply during pregnancy are also important in helping to prevent MMA.

Hog cholera

Also known as classical swine fever (in contrast to African swine fever), hog cholera is a highly contagious acute viral disease, and is frequently fatal.

The disease is spread by animal contact, contaminated urine and faeces or other body secretions. Initial symptoms are a loss of appetite and a high temperature, followed by inflammation of the eyes. The eyes show a discharge causing the eyelids to stick together. There is a severe diarrhoea, trembling and incoordination, and death will often ensue after 4–7 days.

There is no effective treatment. If there is a danger of hog cholera infection, a vaccination programme should be instituted.

Porcine babesiosis (redwater)

This disease is transmitted by the *Rhipicephalus* species of ticks, and particularly affects adult pigs. The prominent features are a high temperature (41–43°C), loss of appetite, weakness, constipation and death within 24–30 hours.

There are effective drugs against babesiosis, but the disease will not occur if access to ticks is prevented.

Salmonellosis

Salmonellosis is another enteric disease, caused by the *Salmonella spp.* of bacteria. Pigs generally are affected around two months of age, and become gaunt, with a high temperature and a foul-smelling diarrhoea. There are usually some deaths in a group of infected pigs.

An outbreak is often triggered-off by a stress condition, particularly heavy worm-infestation. The disease can therefore be prevented by good management and sanitation. Antibiotics and sulpha drugs will aid in the control of the disease.

SMEDI

SMEDI is an acronym for reproductive failure conditions involving stillbirth (S), mummification (M), embryo death (ED) and infertility (I). It is caused by viruses, mainly porcine parvovirus and the enteroviruses.

The symptoms will vary according to when the sow or gilt becomes infected. If infection occurs during the oestrous cycle and at service, the sow will show a regular or irregular return to oestrus, or if only some embryos die she will produce a very small litter. If infection occurs after 35 days of pregnancy, the foetuses die and dry up and are presented at farrowing as 'mummified' foetuses. The condition can cause a serious decrease in sow productivity within a herd.

There is no treatment but effective preventative vaccination programmes are now available. If vaccines cannot be obtained, all gilts and new animals entering the herd should be given access to farrowing house waste 30 days before breeding. This exposes them to the viruses and stimulates immunity.

Swine dysentery

This disease is caused by a large spirochaete, and is manifested by a severe diarrhoea producing reddish-black faeces. Infected pigs rapidly lose weight. The disease is spread by infected dung and can largely be controlled by good hygiene. There are effective antibiotic medications on the market.

Swine influenza

Swine influenza is a highly contagious respiratory disease caused by an influenza virus. It is normally triggered off by a stress, particularly rapid changes in temperature. Although mortality is low, the disease has important economic consequences due to stunting and reduced liveweight gains.

The first sign of the disease is normally a cough, with a high temperature and loss of appetite. The disease spreads rapidly, breathing becomes jerky and the hair coat develops a rough appearance. Secondary infection with bacteria may complicate the condition.

There is no treatment or preventative vaccine available. Infection can be prevented by good management and the avoidance of stress.

Swine pox

Swine pox is a virus disease, and is transmitted either by direct contact or by ecto-parasites such as lice. Small red areas (about 1·25 cm in diameter) appear on the skin around the head, ears and ventral surface, which eventually form scabs.

There is no treatment for swine pox, but although unsightly it rarely causes serious loss and clears up after a short time.

Transmissible gastroenteritis (TGE)

TGE is a virus disease which causes acute diarrhoea, vomiting and early death in young piglets. It also affects older pigs, causing diarrhoea and vomiting, but rarely death.

There is no treatment. Infected pigs can be isolated, or killed and buried. After infection, the whole herd is likely to be immune.

Non-specific diseases

Abscesses

Abscesses can occur as the result of any irritation, inflammation or wound which allows access to bacteria, normally strains of *Staphylococcus* or *Streptococcus*. The body of the pig reacts to the invasion of bacteria, and a pocket of pus is walled off from the body. Abscesses are seen as swellings or lumps, often hot to the touch and they will in time develop a soft area which can be lanced and drained. They may be superficial (Figure 60), or they may form deep within the body, where they can cause lameness, interfere with breathing or swallowing, or may not be discovered until slaughter.

As abscesses are painful and can markedly depress performance and reduce carcass value, every effort should be made to minimise the possible causes in a piggery. Preventative measures include the removal of any sharp or rough objects from pig pens, ensuring the floors are not too rough, especially for baby pigs, making sure that injection equipment is sterilised and providing overall good sanitation.

Abscesses can be treated with antibiotics, but this is not always effective.

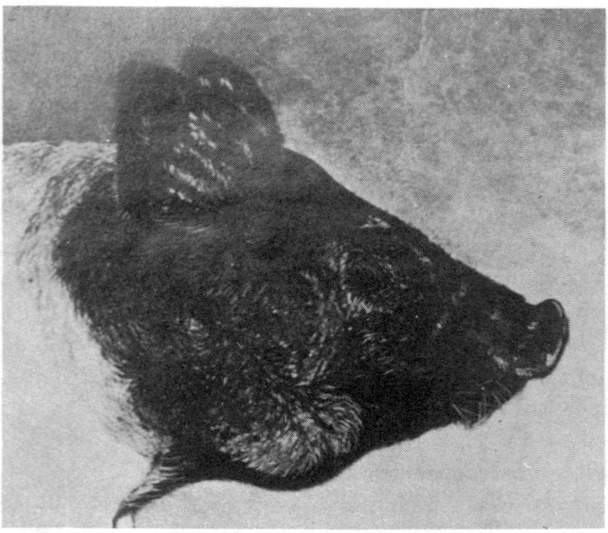

Fig 60 *A painful abscess in the neck region*

Gastric ulcers

Ulcers tend to occur as a response to stress in pigs of all ages, and are particularly prevalent in genetic strains bred for fast growth and a thin covering of back fat. The nature of the ration is also important with a higher incidence of ulcers occurring on finely ground, high-energy concentrate diets.

There may be no specific external symptoms, unless haemorrhaging occurs. Otherwise pigs show lack of appetite, will huddle together and become thin. Mortality varies according to the extent of the ulceration.

There is no specific treatment apart from reducing stress. Changes in the ration involving an increase in fibre levels is often useful in ameliorating the condition.

Intestinal haemorrhage syndrome (Red-gut)

This condition occurs in pigs of 40 kg or over, which are generally found dead. The major post-mortem finding is the presence of quantities of blood in the intestinal tract, without any associated inflammation of the intestine wall.

The cause of this condition has yet to be established, but it is believed that environmental factors may play a part.

Easy diagnosis chart

Symptoms	*Possible disease or condition*
Abortion	Brucellosis. Redwater. Any disease causing a high body temperature.
Agalactia	MMA. Ergot in feed.
Anaemia	Roundworms. Redwater. Iron deficiency. Trypanosomiasis.
Appetite (loss of)	African swine fever. Hog cholera. Redwater. Ulcers. Swine influenza.
Breathing (Distressed/rapid)	African swine fever. Swine influenza.
Constipation	Redwater.
Coughing	Roundworms. Lungworms. Enzootic pneumonia. Swine influenza.
Dehydration	Coccidiosis. Enteric colibacillosis.
Diarrhoea (Scours)	Roundworms. Coccidiosis. Enteric colibaccillosis. Hog cholera. Salmonellosis. Swine dysentry. Transmissible gastro-enteritis.
Discharges	
anus	Anthrax. Intestinal haemorrhage syndrome.
eyes	Hog cholera.
vulva	MMA.

Symptoms	*Possible disease or condition*
Emaciation	Roundworms. Ulcers. Coccidiosis. Tapeworms. Salmonellosis.
Infertility	Brucellosis. SMEDI.
Nervous signs	African swine fever. Hog cholera. Rabies.
Salivation	Foot and mouth disease. Rabies.
Skin conditions	Mange. Lice. Fleas. Erysipelas. Swine pox. Greasy pig disease. African swine fever.
Sudden death	Anthrax. Acute trypanosomiasis. Acute erysipelas. Intestinal haemorrhage syndrome.
Swellings/blisters	
face	Foot and mouth disease. Abscesses.
body	Abscesses. Anthrax. Gut oedema.
feet	Foot and mouth disease.
udder	MMA.
Temperature elevated	All bacterial, protozoal and viral diseases.
Urine (discoloured)	Cystitis. Redwater.

8 Practical management

General considerations

Without doubt, management has the overriding influence on the profitability or otherwise of a pig herd. The producer must strive to integrate all the considerations relating to the well-being and productivity of his animals with those relating to cost-effectiveness. Success in this regard invariably depends on a relatively high degree of attention to detail.

Some important factors in a piggery are:

- *Stockmanship* This concerns the personal relationship and rapport between a stockman and his pigs. Pigs are very responsive to quiet, calm conditions. The good stockman is firm, but understands the needs of his pigs. He will always be on the watch for any irregularities in the behaviour or condition of a pig and thus learns to anticipate problems before they arise.
- *Pig handling* This is an adjunct of good stockmanship, and should be carried out in as calm a way as possible. It is particularly important to minimise the upset to pigs when they are caught or driven. When catching a suckling pig, it should be approached from behind and lifted suddenly by one or both hind legs above the knee-joint or hock. It can then be held round its shoulder. Bigger pigs can be caught by grasping with both hands behind the shoulder of the pig, lifting it off the ground and then holding it firmly against the body. If pigs have to be driven, it is essential to remember that they cannot be led by the head (like a cow, for example), but must be driven from behind. The natural instinct of the pig is to head for any gap and the skill is therefore to make maximum use of any walls, solid hurdles or pig boards to ensure that the only gap is in the direction you wish the pig(s) to proceed.

  Occasionally, it is necessary to restrain a sow or boar for treatment purposes. The best method is to make a slip-knot in a two-metre length of rope and allow a sufficiently large noose around the pig's mouth, encircling the snout close up against the juncture of the upper and lower jaws. When in place draw the noose tight. If necessary, twist the surplus rope round a railing or pole, but normally a man holding firmly at about 35° to the horizontal will be able to hold the pig, which will always pull against the pressure and stand firm for any treatment. For quick release, do not pull the rope right through when completing the knot, but leave it in a loop. When the loose end is pulled, the knot will fall apart.

- *Cleanliness* The importance of good hygiene in a piggery cannot be over-emphasised. Not only does it help to reduce the incidence of disease in the pigs, but it also has a beneficial effect on staff morale by improving the working environment. Any cleaning programme that is designed should allow for the periodic emptying and resting of each house. For the farrowing house, this should consist of a week's rest in between each batch of farrowing sows. For fattening pens, a five-day break after each group of fatteners is adequate. Immediately after it is empty, each pen should be thoroughly scrubbed and cleaned, washed and soaked in disinfectant and allowed to remain dry for the remainder of the period that it is empty.

Boars

Although it is often said that 'the boar is half the herd', the management of boars is frequently one of the most neglected aspects of a pig enterprise.

The young boar

When young boars are brought into the herd, care and attention whilst they settle into their new surroundings are paramount. Boars should be purchased well in advance of their expected use, and housed on their own for a month. They should not, however, be isolated from sows and gilts. On arrival, they should be dosed for worms and sprayed/dipped against mange infection. It is recommended that the young boar is exercised daily. This has the added advantage of allowing him to get used to his stockman, and the sights and smells of the piggery. The boar should be fed to provide for continued growth, but should not be allowed to become fat and sluggish. Two to two-and-a-half kg of a standard sow and weaner meal each day is usually adequate for this purpose.

A boar should be ready for service at seven-and-a-half to eight months of age. In the first instance, he should be introduced to a small sow who is well on heat and standing firm. This will prevent him becoming frustrated and allow for a successful first service and a favourable sexual experience. At no time should a young boar be introduced to a group of females, as the bullying that may ensue can adversely affect the boar's libido for some time afterwards. For the first few months of his breeding life, the boar should not be overworked and a maximum of two services per week should be the rule.

Mature boars

Condition

Maintaining a boar in the lean, hard condition which is ideal for service requires good stockmanship. A balanced ration should be fed at a level which prevents the boar becoming either overfat or too thin.

Environment

Research has shown that isolation of mature boars (i.e. older than 10 months) from female pigs can severely reduce their level of sexual behaviour. Thus, boars should be housed within sight and sound of sows and gilts so that they obtain the necessary stimulation from the presence of female pigs.

Under tropical conditions, it is imperative that boars are protected as much as possible from high temperatures, otherwise they will suffer from depressed libido and sperm production. This can be achieved by good design (see Chapter 6). In hot climates spraying with water, either by hand or by sprinklers which are timed to operate intermittently and allowing access to wallows can all be beneficial in reducing heat stress and sunburn. Conducting matings either early in the day or in the evening when temperatures are normally lower is also important.

Mating conditions

Emphasis has already been placed on the necessity for a specifically designed service pen (see Figure 61 and Chapter 6). Confirmation of the benefits of a special service pen has been provided by a recent Australian study, where boars mating in their own pens displayed lower levels of sexual behaviour than boars mating in the service pen. This resulted in a lower percentage of gilts being mated in the boar pen.

Boar-to-sow ratio

To avoid overworking boars the ratio of sows to boars should not be too large. The standard recommendation is one boar to 20 sows and gilts. However, this must be related to peaks of service activity. In units where boar numbers are low, it may be beneficial to allocate two days each week instead of one for weaning sows so as to spread the work-load of the boars.

Frequency of use

There is a great variation between boars in their ability to work consistently, but in general terms overworking a boar will reduce the quality

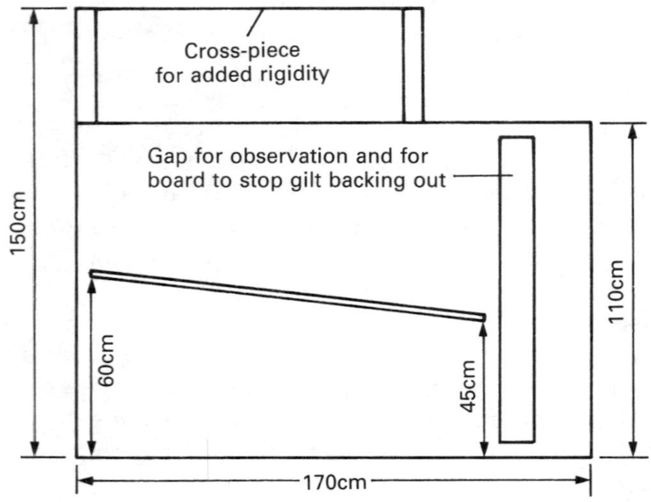

Fig 61 *A service crate* (After Manners, 1955, in *Journal of the Ministry of Agriculture (England and Wales)*, Vol. 59, p. 419)

and quantity of the sperm produced, leading to small litters and increased number of sows returning to service.

By ten months of age, a well-managed boar should be able to cope with five to six services per week and still maintain good fertility levels. On the other hand, a boar should not be under-worked, and intervals of over 14 days will likely lead to reduced litter-sizes because the semen becomes less fertile.

Lameness

Surveys from many countries have shown that some 20 per cent of boars are culled for foot problems and lameness. Prevention is better than cure, and care should be taken to ensure floors are not rough or dirty. New concrete is particularly damaging to skin and feet because of its extreme alkalinity.

Supplementary biotin (a vitamin) in the food (see Chapter 5) and the use of footbaths containing either 5 per cent formalin or 5 per cent copper sulphate solution can help to harden the claw horn.

Sows

The young sow (gilt)

Normally, sows are culled after their fourth lactation. This means that sows with their first litter contribute some 25 per cent to herd productivity, and highlights the need to optimise their performance.

Gilts which are selected on the farm should be moved out of fattening pens at five months of age into groups in well-bedded pens or yards, or even outside paddocks. They should be fed to continue growth without becoming overfat. Two to two-and-a-half kg of a standard sow and weaner meal a day is usually adequate.

If gilts are bought into the herd, special care and attention and a period of 'integration' into the herd are essential. Ideally, they should be purchased in batches at least six weeks before they are due for service. They should be examined carefully to see that they are not damaged in any way and every encouragement given to start them eating normally in their new environment. The gilts can be housed in pens or yards, and given access to dung and waste from the farrowing house, allowing them to build up immunity to infertility viruses and other diseases. Thereafter, they can be gradually integrated into the main gilt/sow herd.

The major objective of gilt-management should be to induce all replacement gilts to reach puberty as soon as possible after selection. This will allow the following objectives to be achieved:

- disposal of gilts which are not showing any breeding activity at an early stage;
- access to a pool of young, sexually-active gilts;
- gilts to be at second heat or more at first mating, thereby increasing first litter-size (see Chapter 2).

In order to help stimulate the onset of puberty, a boar should be introduced to the gilts for 10 to 15 minutes every day. The boar used for this purpose should be at least 9 to 10 months of age and actively working so that he is producing enough of the necessary odours (pheromones) to have a stimulatory effect on the gilts. Gilts are most sensitive to boar stimulation at 145 to 170 days of age.

If gilts show reluctance to come on heat for the first time, other techniques which can sometimes be beneficial are:

- mixing them with strange gilts;
- moving them into new accommodation;
- taking the gilts for a ride in a truck or trailer.

Gilts of exotic breeds should be over 120 kg bodyweight when served and in good active condition. *Ad lib* feeding for 14 to 21 days prior to service helps stimulate the number of eggs and consequently litter-size. Care must be exercised that the boar used on gilts is not too heavy, otherwise he may cause damage. If the boar is too heavy, a service crate (Figure 61) can be useful. The sow stands in the crate and the boar takes some of his own weight on his front legs, which rest on planks. However, successful service using a service crate is not always easy to achieve, and it is generally best to try and avoid too much disparity in weight between the boar and the gilt.

Lifetime productivity

Although much emphasis is placed on output of the sow in terms of numbers of piglets reared per litter, the ultimate measure of her productivity and profitability is the number of healthy weaners reared in her lifetime. Management strategy must therefore be aimed at maximising this output within the constraints of each enterprise.

However, before considering the separate phases in the breeding cycle of a sow, namely farrowing, lactation, weaning, mating and pregnancy, it is essential to realise that they cannot be dealt with in isolation. Each phase will impact on the other and with feeding, for example, there can be large carry-over effects from one phase to the next.

Views on the best ways of feeding sows have varied over the years, but it is now generally agreed that the best guideline is to minimise bodyweight fluctuations over the entire reproductive cycle. This means relatively low levels of feeding during pregnancy and high (*ad lib*) levels during lactation. Not only does this prevent the sow becoming too fat and heavy at farrowing, which minimises farrowing problems, but it also keeps her from becoming too thin at weaning and causing delayed oestrus and rebreeding. At the same time, allowance should be made for continued growth and weight-gain for at least the first three to four reproductive cycles. In this respect, it is often recommended that a sow should gain some 12 to 15 kg of bodyweight from weaning to weaning for the first four parities (see Figure 62).

Recent work at Edinburgh University by Professor Whittemore and his colleagues has indicated that, particularly for modern leaner sows, the state of the sow's body-fat reserves is likely to be more important

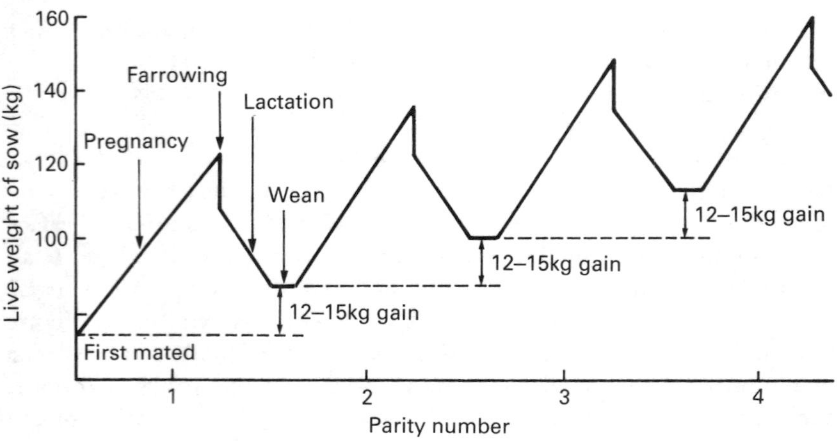

Fig 62 *Recommended bodyweight increases in sows over successive parities*

for reproduction than weight gain. A 160 kg sow, for instance, may be a small-framed animal carrying a large amount of fat, or a much larger-framed animal in very poor condition. As a consequence, a sow condition scoring technique has been developed and is widely used as a means of monitoring the body reserves of sows at strategic points throughout the breeding cycle. Sows are scored on the basis of a combination of visual appraisal and physical feel at various points such as the pin bones, tail setting, loin area, backbone and ribs. A simplified description of the sow condition scoring system is given in Table 8.1. By aiming to achieve certain condition scores at set points throughout the breeding cycle, this provides a very useful guide to management and feed requirements.

Table 8.1 The sow condition scoring system (after Agricultural Development and Advisory Service, UK)

Score	Description	Assessment of fat cover over: backbone transverse spinal processes hips
0	Emaciated	Exposed, no cover on bones
1	Poor	Bones prominent, little cover
2	Moderate	Bones easily felt without palm pressure
3	Good	Bones only felt with firm palm pressure
4	Fat	Bones cannot be felt with firm palm pressure
5	Grossly fat	Further deposition of fat impossible

The change in approach from feeding according to set requirements to the more flexible system of feeding according to the body condition of the sow is particularly appropriate in the tropics, as it allows for more effective use of less conventional foods of lower nutrient density. The adequacy of these foodstuffs can be assessed by the responses in sow condition particularly when they are fed to dry sows.

Farrowing

Farrowing, or parturition, is undoubtedly one of the most critical stages in the whole reproductive cycle. Problems at this stage lead very quickly to high mortality rates and reduced efficiency of both sows and piglets. The skill of the stockman is to learn to recognise what constitutes a normal farrowing so that any departures from normal can be detected and corrected.

Two to three weeks prior to farrowing, the sow should be dewormed to minimise the possibility of the sow passing worms to the new-born litter. She should be moved to her farrowing quarters a week before she is due to farrow. It is advisable that a sow should be washed with soapy water before entering the farrowing pens as this removes any dung and also sprayed to kill any mites and lice on her body (see Chapter 7).

Constipation in the sow must be avoided at all costs as this interferes with the farrowing process. If there is a tendency in the herd to constipation, the diet should be supplemented with bran, any greenstuff or extra fat, for the last week of pregnancy.

There are two sure signs that farrowing is about to commence, namely:

- the sow will show increased restlessness (in direct contrast with the peaceful animal of late pregnancy), and will start to make a nest by rearranging her bedding.
- milk can be withdrawn from the udder about 12 hours prior to the start of farrowing.

  Although not essential, it is recommended that for sows farrowing under intensive confinement conditions, each farrowing should be supervised by a stockman. The best stockman in the unit should be assigned to this task. Clearly, this is not practicable under extensive conditions when sows are generally better able to farrow unaided.

Two major problems encountered during farrowing are:

- the process may become painful, particularly for gilts if the piglets are relatively large. This can result in the dam savaging her litter. Should this be observed, all piglets should be removed until farrowing is completed and then reintroduced slowly under supervision. Exercising the sow can also help, but it may be necessary to muzzle (part of the leg of an old gumboot is very useful for this purpose) or tranquilise the sow.
- *Delayed farrowing* After farrowing has started, any time-gap of greater than 30 minutes between presentation of piglets probably means that a piglet has stuck in the birth canal. In this case, the stockman should disinfect and soap his arm and then gently work it into the birth passage to release the stuck piglet. Because the sow will often stop straining in this situation, a simultaneous injection of oxytocin will usually get contractions going again and farrowing will then proceed normally. The only certain sign that farrowing has been completed is the passage of the afterbirth.

Stillbirths

Surveys in Europe have shown that stillbirth rates can vary from 1 per cent to more than 12 per cent of all piglets born, which means that, in some cases, it is the single most important cause of piglet mortality and seriously depresses productivity.

It is believed that over 70 per cent of stillbirths are caused by asphyxiation or weakening during the birth process and that some 70 per cent of these occur in the last 30 per cent of piglets born. This is due to the design and length of the sow's uterus (see Chapter 2), which means that piglets from the far end have to travel some 0·3 metres after their umbilical cord has ruptured. The intact cord supplies the piglet with oxygen from its mother's blood; once the cord is broken, the piglet will start gasping for oxygen. It then has five minutes to get to the outside before it will stop breathing.

Many factors can cause delays in the farrowing process and thereby increase the incidence of stillborn piglets, namely:

- *Large litters* Because the whole process will take longer, the last piglets to be born have a higher risk of being suffocated.
- *Very small litters* These are liable to provide insufficient stimulus to the sow to start effective uterine contractions.
- *Older sows* After five or six litters, uterine tone declines, resulting in a less efficient birth process.
- *Genetic make-up of the sow* Variations can occur between families in the efficiency of farrowing. If a strain of pigs is identified which have a regularly higher incidence of stillbirths they should be culled.
- *Nutrition* Sows should not be overfat or, alternatively, too thin with insufficient reserves of energy.
- *High ambient temperatures* Sows will tend to tire more quickly at high environmental temperatures.
- *Piglet diseases* Dead or mummified piglets can slow down the birth process leading to a greater risk to the live pigs.

If stillbirths are a problem in the herd, every attempt should be made to minimise the effects of any or all of the above factors.

The newborn piglet

The crucial point in relation to managing piglets to ensure high survival is to remember that the piglet is born with very limited supplies of energy (see Chapter 2). This is often exacerbated by the stresses of the birth process.

A healthy, good-size piglet will get to its feet within minutes after birth and instinctively try to reach the sow's udder. On average, such piglets obtain their first successful suckle of colostrum within 45 minutes after birth. Once this first suckle of high-energy colostrum has been achieved, the piglet is better able to continue to commandeer a teat. However, weaker and lighter piglets (less than 1·0 kg at birth for exotic breeds), are less able to reach the udder, are generally less competitive if they do reach it and are the most liable to die if they are not assisted.

Apart from air, colostrum is the most important substance a piglet takes in during the first few hours of its life, because it is a source of both essential energy and antibodies. Failure to obtain colostrum will invariably result in death of the piglets.

Management techniques which can assist in piglet survival are:

- *Maintaining temperatures* Ensuring the piglet is not exposed to cold or draughts, because if it is it has to utilise energy to keep warm.
- *Cross-fostering* This is a well-proven and widely-adopted technique. When sows farrow within a few hours of each other, litters are evened-up by transferring piglets from sows with large litters to those with smaller litters, thereby increasing the survival chances of the smaller piglets in the original larger litters. When cross-fostering, it is advisable to transfer the bigger piglets on to the foster mother. Care must also be taken that when piglets are transferred on to different sows, the teats are not too big for them to be able to suckle successfully.
- *Split suckling* After the larger piglets in the litter have had a good suckle of colostrum (approximately 1 hour at the udder), they are shut away in a warm box for the next two hours. This gives the remaining smaller piglets access to the udder to get their share of colostrum. This can be repeated twice in the first 18 hours.
- *Supplementation* The supply of colostrum to small and weak piglets can be augmented by supplying them with colostrum which has been stored. Colostrum can be obtained by milking sows before they farrow, or during farrowing, and storing it in the deep freeze. This can then be given to needy piglets via a bottle, or more effectively, via a stomach tube. The latter consists of a soft tube of around 25 cm in length which is inserted down the throat directly into the stomach.
- *Extra fat in the sow diet* The addition of some supplementary fat in the diet of sows for seven to 10 days before farrowing can increase the fat, and therefore energy, content of the sow's milk and may slightly increase piglet energy reserves and vitality.

Other routine operations which it is advisable to carry out at this stage are:

- navel cords should be immersed in a dilute solution of iodine within 12 hours of birth. Any individual navel cords which are very long should be trimmed to ± 5 cm before immersion in the iodine solution. This prevents infection entering through the navel and causing joint-ill, a condition causing swollen joints. Piglets with the condition will not thrive and may die.
- eye-teeth should be clipped within 24 hours to remove the points. This process helps to prevent teat and udder injuries.
- within three days after farrowing, housed piglets should receive supplementary iron to prevent anaemia. This can be given by injection, which is preferable. If injectable compounds are not available, soil enriched with ferrous sulphate should be placed in the pen.
- ear-notching, or other systems of identification, can also be carried out at this stage.

The lactating sow

Contented sows make the best mothers, and management should be aimed at keeping sows as comfortable as possible.

Directly after farrowing, the sow should be examined for any disease problems, particularly MMA (see Chapter 7), and if symptoms are seen, she should be treated immediately.

As the sow can lose up to eight litres of body fluids from the uterus and associated tissues over a four-to-six hour period at farrowing, it is essential that she has access to plenty of fresh water. Failure to provide water at this stage can lead to agalactia.

If the sow is not too fat at farrowing, the lactating sow should be fed at a rate designed to minimise her bodyweight-loss during lactation. Unless she has a very small litter, this generally means feeding as much as she will eat, twice daily. If a sow continues to show significant body-mass loss in spite of feeding to appetite, intake can be stimulated by feeding more frequently. Also, giving wet rather than dry food, particularly in hot climates, helps to further stimulate appetite.

Because of the heavy demands on the sow during lactation, care must be taken to ensure that environmental temperatures at sow level are within her thermoneutral zone (see Chapter 2).

In tropical regions, it is not recommended that the lactation period is any shorter than five weeks, and in certain cases, it may be more cost-effective for it to be longer. Although this contrasts with the situation in developed countries, where four- and three-week weanings are often practised, the lack of quality creep feed for the young pig in developing countries means that attempts to wean earlier than five weeks have generally been unsuccessful.

Creep feeding

The name of creep feeding has developed because the feed has traditionally been given in the piglets' nest or creep where it is not accessible to the sow. The milk supply of the sow reaches a peak about three weeks after farrowing and thereafter declines. Therefore after this stage it is necessary to provide the piglets with some solid food to make up the shortfall.

Other benefits that accrue from the practice of creep feeding are:

- enhanced growth in piglets when compared with those that only have access to sow's milk.
- the provision of creep feed helps the piglet's digestive system to adjust to the change from a milk to a solid diet. This reduces the chances of digestive upsets and checks in growth rate at weaning.
- it helps reduce the drain of nutrients from the sow, minimises her

bodymass loss in later lactation and leaves her better prepared for rebreeding.

- it attracts the piglets away from the vicinity of the sow and reduces the chances of the piglets being crushed by the sow.

Creep feed needs to be specially formulated to be highly digestible and palatable. It is required to meet the nutritional needs of young pigs prior to weaning and afterwards up to eight weeks of age. Ideally, therefore, it should be based on skim-milk powder with added unsaturated fats and good quality non-milk proteins. Because milk products tend to be both scarce and expensive in the tropics, and often restricted to human use, creep feeds tend to be of relatively low quality in these areas. Nevertheless, simpler diets based on maize and soyabean meal can still be very effective.

Creep feed should be first offered to piglets at about seven days of age in very small amounts in order to accustom the piglets to consumption of some solid food. It can then be increased according to appetite up to weaning. The feed should always be offered fresh on a 'little and often' basis and a separate supply of fresh water for the piglets will help stimulate food intake.

Weaning the sow

At weaning, the aim must be to dry the sow off and stimulate her to exhibit oestrus and conceive again as soon as possible. At the same time, she should be induced to release a large number of eggs at ovulation as the first step towards a large litter at the next farrowing.

Under normal conditions the removal of the sow from her litter and the cessation of the suckling stimulus triggers ovarian activity, and oestrus will occur within four to seven days. Management aids which can be used to help ensure that oestrus is not delayed are:

- moving the sow into a house where she can hear, smell and have contact with a mature boar.
- housing her in the presence of other newly-weaned sows because other sows coming into oestrus will also help the stimulation process. Newly-weaned sows should not be mixed unless/or until they are used to each other.
- sows should be able to exercise. The cubicle house design (see Chapter 6) is ideal for this purpose.
- feeding the sows at a high level after weaning until oestrus occurs. This is particularly beneficial for young sows after their first litter or for sows that are thin.

Mating management

The crucial points in effective mating management are to recognise the onset of oestrus and then to serve at the correct time. By ensuring sows are served during the period of peak fertility (see Figure 37), this will reduce the likelihood of sows returning to service or producing small litters.

Characteristic signs of oestrus under practical conditions are:

- swelling and reddening of the vulva (not always obvious);
- confined sows will spend less time lying down;
- sows will be more alert and the ears will stand up in prick-eared breeds;
- sows will allow themselves to be mounted by the boar or other sows;
- sows will give a characteristic 'grunt';
- sows will exhibit the 'standing reflex', i.e. they will adopt an immobile posture when pressure is applied to the back.

Ideally, sows should be served in a specially-designed service area (see Chapter 6), and each service should be quietly supervised by a stockman to ensure that an effective mating has occurred. In the tropics, sows can quite easily be served outside under a tree and this is generally preferable to serving in the boar pen if a special service area is not available. Care must be taken that sows are served during the cooler time of day.

Under careful management, a single mating is perfectly capable of giving rise to good conception rates and large litters. However, it is generally safer to serve sows twice, once approximately 12 hours after the sow is first seen on standing heat, and the second time a further 12 hours thereafter. In some units a triple-mating procedure at 12-hour intervals is used. This can be considered if unsatisfactory results are obtained with a double-mating system.

The pregnancy period

The day after service, the amount of food fed to sows should be reduced, because high feed-levels during this stage are associated with increased losses of embryos. Other steps which can be taken to minimise embryo-losses are to prevent excessive environmental temperature affecting the sows and to prevent any undue stress on the sow during early pregnancy.

All sows should be checked during early pregnancy to detect for any that have returned to oestrus so that they may be served again. If available, a pregnancy-detecting instrument can be used to confirm pregnancy. There is a variety of instruments on the market. The majority of these use a sonar principle to detect the increased blood flow to the uterus and foetal heart beats which occur in early pregnancy. Failure to check for empty sows can be costly, because by the time they are finally discovered they should have been due to farrow.

The pregnancy period provides an ideal opportunity for the feeding of bulkier feedstuffs. Not only is the sow's appetite well above her requirement for nutrients given in concentrated form during this period, but the 'fill' effect of bulky foods keeps her more content. As long as the ration is balanced and sow condition is maintained, considerable savings in food costs can be achieved. Once-a-day feeding of dry sows will also save costs.

Extensive sow systems

The types of sow used on more extensive systems, be they indigenous, indigenous × exotic, or exotic hybrids, are generally hardier and exhibit better mothering ability than exotic sows bred for confinement conditions. Nevertheless, sows run on extensive systems respond to good management and stockmanship. Some important aspects are as follows:

- provision of adequate deep shade and wallows;
- siting farrowing arks or nests sufficiently far apart to allow sows to establish their territory at farrowing time;
- provision of bedding in the farrowing arks for nest-making purposes;
- ensuring that stockmen gain the confidence of sows so that they can be handled easily during critical periods such as farrowing time.
- ensuring that bulky foods are spread over a large area at feeding time so that all sows have the opportunity to gain access to the food. If cassava roots are placed in a heap, for example, dominant sows will often guard it and prevent the more timid sows from eating.
- provision of an adequate boar: sow ratio.
- adoption of a rigorous de-worming programme, unless regular rotation of sows on to new land is practised.

Weaners

Stresses at weaning

The three-week period immediately after weaning is a critical one for the young pig, because it is simultaneously confronted with a number of stresses, namely:

- the psychological trauma of separation from its dam;
- the social stress of mixing with pigs from other litters;
- the stress of settling into the strange environment of a new pen;
- the major change in diet associated with the loss of the sow's milk;
- the loss of the supply of immunoglobulins and immune protection provided by the sow's milk.

Objectives at weaning

The objectives must be to achieve the transitional process of weaning without any growth checks or mortality and to obtain accelerating growth of pigs over the period. This is important because the young pig's potential for efficient growth at this time is high, and any factor which reduces growth rate can prove costly. However, the task is complicated by the fact that the major change in diet often leads to digestive scours, which in turn provides an ideal environment for multiplication of *E. coli* and problems of gastroenteritis and bowel oedema (see Chapter 7).

Management considerations

Good management over the weaning period can minimise the stress and result in uninterrupted growth and low mortality. Some important considerations are:

- immediately before weaning, pigs should be handled and disturbed as little as possible. Any essential management tasks should be carried out at least two weeks before weaning.
- piglets should be encouraged to consume as much creep feed as possible before weaning and should continue to be fed creep feed for at least two weeks after weaning.
- *ad lib* feeding is recommended in order to capitalise on the high growth potential of the young pig. However, as soon as any scours are seen, feed intake should be restricted. If the problem persists, consideration should be given to increasing roughage levels and reducing the protein content of the diet, but this must be done with due consideration as antibiotics can easily be misused. The continual use of an antibiotic as a prophylactic can lead to the build-up of resistance in the bacterial population to the effect of the drug. This renders the antibiotic relatively ineffective in the face of a major disease outbreak. To avoid the build-up of resistance, care must be taken that antibiotics are always used at the correct dosage rate, and that some antibiotics are reserved only for therapeutic use. Antibiotics can also be included in the diet to help prevent scours.
- at weaning, batches of pigs should be made up by selecting them from different litters according to bodyweight. Excessive fighting after mixing can be avoided by supervision for a few hours or, alternatively, by smearing the pigs with oil. Pigs can be retained in their litter groups, which overcomes any problems of fighting, but results in batches of different size and tends to give rise to uneven growth rates.
- excessive variations in ambient temperature within the house should be minimised (see Chapter 6). Pens should never be overcrowded so that pigs cannot spread out to keep cool. Wet floors or wallows should be provided if excessive heat is a problem.
- ideally all piglets should be in good condition at weaning, so that they possess some fat reserves to help combat stress. Poor litters should be

allowed to remain with the sow for longer than five weeks. A technique of 'split' weaning can also be beneficial, whereby the large piglets are weaned first and the small ones are left with the sow for a further week.

- fresh water must be available at all times to the piglets. This encourages higher intakes of creep feed.

Growing and finishing pigs

Objectives

By eight-to-nine weeks of age, the growing pig is over the stresses of weaning, and its digestive system will be competent to deal with a range of protein and energy sources. Some 80 per cent of the food used in a pig unit is consumed by the growing and finishing pigs, therefore the efficiency of food utilisation during this phase is a crucial factor affecting profitability.

Clearly the management system which is adopted must relate to the specific objectives of each unit and these may range from home consumption of as cheap a carcass as possible to the production of sophisticated bacon products. These considerations dictate the type of pig to be produced. Feeding and management must then be geared to optimise performance (see the discussion in Chapter 2). The small-scale rural producer, for example, will attempt to maximise the use of cheaper, lower-quality feedstuffs. It must be remembered, however, that the growing pig has only a limited ability to digest and utilise fibre in the diet and too much bulky food may depress growth to such an extent as to render its inclusion uneconomic. Under commercial conditions, different priorities may be given to factors such as food-conversion efficiency and food cost per pig, growth rate, carcass leanness and grading according to the relative economic advantages they confer in the whole production unit.

Traditionally, the main categories of slaughter pigs in commercial industries have been:

- porkers, which are slaughtered at liveweights up to 65 kg;
- baconers, which are slaughtered from 70 to 90 kg liveweight;
- heavy hogs, which are those pigs slaughtered at liveweights between 90 kg and 140 kg

In general terms, the heavier the pig is at slaughter, the cheaper is each kilogram of carcass meat that is produced.

Feeding systems

The major decisions to be taken in relation to the feeding system to be adopted are as follows:

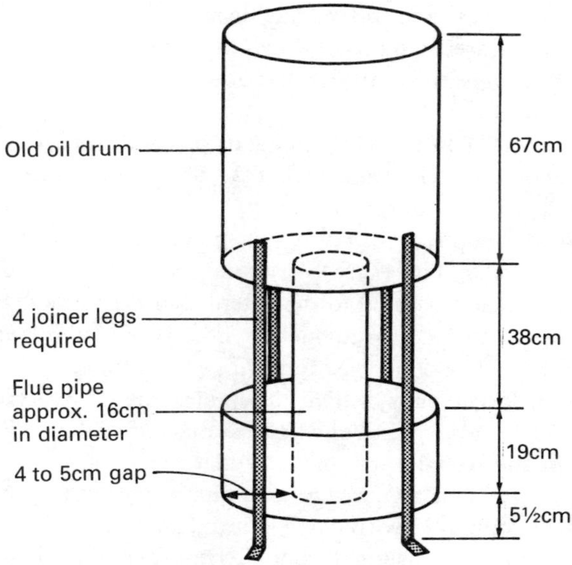

Fig 63 *A self-feed hopper made from an old oil-drum*

- *Restricted or* ad lib *feeding* In the past, all fattening pigs have tended to be fed on restricted feeding regimes to avoid producing carcasses that are excessively fat. However, modern exotic strains of pig have the ability to grow through to slaughter at younger ages and without getting too fat, and can therefore be fed *ad lib*. Thus the decision on whether it is necessary to restrict feed intake will depend on the circumstances of each unit.

  The advantages of *ad lib* feeding are that it saves labour, gives faster growth-rates and a larger carcass-weight at a given age. In general, it will result in more contented pigs and avoids problems of competition at the feed trough. Care must be taken that *ad lib* feeders are designed to operate efficiently in order to minimise waste. A suitable, relatively cheap and efficient feeder which has been developed in Africa is shown in Figure 63. If it is necessary to limit nutrient intake, dietary fibre levels can be increased.

  Obviously, various combinations of *ad lib* and restricted feeding periods can be used in order to yield the most desirable result. Feeding *ad lib* up to 50 kg body weight and then restricting intake until slaughter, for instance, is a system in common use.
- *Feeding in a trough or on the floor* Feeding on the floor has the advantages of more economical use of floor space, and the reduction of bullying and competition effects. However, although feeds should be restricted to what is consumed in 20 minutes, it leads to more waste, a greater

problem with dust and, if scouring occurs, a greater problem in transmission of disease than occurs when pigs are fed in the trough. If pigs are fed in troughs, it is critical that adequate trough space is provided (see Chapter 6).

Frequency of feeding in troughs can also influence the efficiency of feed usage. Research has shown that by feeding twice a day rather than once, small improvements are achieved in growth rate, food required to slaughter and killing-out percentage (see below).

- *Wet or dry feeding* There are two main types of wet feeding systems. Firstly, the food is diluted to the extent that it can be delivered to the pens by pipeline or continuous trough (river system) and secondly, water is added to food in the trough. The first system is really appropriate only in larger units, where the mixing and delivery can be semi-automated. Trials in Zimbabwe have shown that the addition of water to food in the trough, at a meal-to-water ratio of 1:1 or 1:2 can save some 5 per cent in food usage and decrease time to slaughter by 10 days when compared with dry feeding. If food is fed in the wet form, it must be totally consumed each day in order to avoid spoilage.
- *Meal or pellets* Pelleted feed has been consistently shown to be superior to meal by some 5 per cent in promoting liveweight gain and food-conversion efficiency. However, these improvements in performance have to be compared with the extra cost of pellets. Moreover pelleting equipment is not always available in the tropics.
- *The handling of bulky food* A system which has proved successful in small- and medium-scale operations is the separate feeding of the basal feed and the protein concentrate. In systems in Madagascar, for example, fresh cassava roots are dug twice a week and fed directly to the pigs to appetite. Protein concentrate is then fed daily according to a fixed rate (500 g for weaners up to 50 kg liveweight, 750 g for finishers above 50 kg). Such 'separate feeding' systems maximise the added value from farm production. Although they may be sub-optimal in terms of nutrient balance, they optimise the economic efficiency of production.

Water

The importance of water to the growing pig in hot climates cannot be over-emphasised. The art of successful fattening is to stimulate maximum feed intakes. The absence of a regular supply of clean, fresh water is often the first factor that limits food intake.

Sex

If pigs are reared intensively and achieve fast growth rates, there is little justification for castration of male pigs (see Chapter 2). It is therefore recommended that, except for small units where it is impractical, the sexes should be segregated for fattening purposes. This allows for the provision of different feed requirements and also prevents the early-maturing boars from continually riding the gilts. However, on more extensive systems of production where males grow more slowly it may be necessary to castrate males to avoid the occurrence of 'boar taint' in the carcass at slaughter.

Vices

Tail-biting is the most common vice, but other forms of cannibalism such as ear-biting and vulval-biting can occur. Once blood is drawn, very serious wounds can develop which often lead to secondary infections.

Vices mainly occur in pigs that are kept in intensive confinement conditions and are generally considered to be a response to boredom. Many factors will trigger off tail-biting, but in hot conditions, over-stocking, lack of bedding, poor ventilation and lack of salt would appear to be the most important.

Provision of a small amount of bedding will invariably prevent tail-biting. If this is not available, the hanging of a chain from the roof as a toy can be beneficial.

Killing-out percentage

The killing-out percentage, which is the ratio of carcass mass to livemass, can have a significant effect on profitability. Pigs fed bulkier foods of lower nutrient and energy density will develop a larger gut capacity in proportion to body size and therefore have a markedly lower killing-out percentage.

Culling of sows

In order to maintain overall productivity in the herd, it is important to have a culling policy so that sows are removed at the correct time. In the majority of cases the reasons for culling sows will be obvious, e.g. lameness, other injury, farrowing problems, poor litter-size, poor mothering ability and low fertility. However, the performance of a sow which regularly produces a good litter will eventually start to decline with age, probably around her tenth litter. If she is producing well, a good guide is to allow her to remain in the herd until her performance falls below the average of the gilts in the herd. At the same time it is important to have a supply of pregnant gilts available to replace sows that need to be culled.

The once-bred gilt

Mention should be made of a relatively new system of producing weaners which could have applicability in the tropics. This involves the serving of gilts at an early age and then slaughtering them after they have weaned their first litter. The carcass of the gilt is very similar to that of the heavy hog. The system exploits the phenomenon known as pregnancy anabolism, whereby the deposition of maternal tissue during gestation is very efficient in terms of food utilisation. Approximately 50 per cent of the total weight gain during pregnancy will consist of body gain, and the other 50 per cent will comprise the products of conception. It should be considered as an adjunct to a conventional sow-based system. Some older sows will generally be required in the herd to maintain immunity against reproductive diseases. Nevertheless, it could prove a useful way of producing cheaper meat in a pig enterprise.

Recording

In order to monitor both the technical and economic efficiency of a pig enterprise, it is absolutely essential to keep some basic records. These should cover all aspects of the enterprise, so that evaluations can be made of boar performance, sow productivity, weaner growth and grower/finisher efficiency. By looking at the recent history of the enterprise, each phase of production can be critically examined and weaknesses pinpointed. It is then possible, based on facts, to plan to rectify these weaknesses.

Many types of recording systems are available, but Tables 8.2 – 8.6 provide some examples of simple recording sheets which can be used. It is important that provision is made for day-to-day recording, and these records can then be used for compiling individual animal and herd performance sheets for assessment.

For breeding, the first essential is to keep an accurate service record of all sows and gilts in the piggery (Table 8.2). Farrowing and lactation performance details can then be kept on individual sow record cards (Table 8.3). All this information is available for transfer on to a sow lifetime record card (Table 8.5), which enables an assessment of her contribution to the herd performance. This information can also be used to compile a record on individual boar performance (e.g. Table 8.4). As major differences in boar performance become evident, the lower-performance boars can be removed from the herd.

These data can then be combined, together with details of pigs sold, to give an overall herd performance report (Table 8.6). This allows the producer to monitor the performance of his herd on a monthly and six-monthly basis.

Table 8.2 A sow service record sheet

Sow ear no.	Date weaned	Parity	Services						Date due	Date farrowed	No. born		Weaned		Weaning age (days)
			Date	Boar	Date	Boar	Date	Boar			Alive	Dead	No.	Date	

Table 8.3 An individual sow record card

Sow no.______	**Date served**______	**Boar no.**______

Due to farrow ______ Farrowed ______
Total born ______ Born alive ______
Pigs fostered off ______ Pigs fostered on ______
Date iron injection ______
No. pigs weaned ______ Date weaned ______

Total litter wt weaned ______
Average/weaning wt ______

Remarks ______

Table 8.4 A boar performance record sheet

Boar no.______ **Date of birth**______ **Breed**______

Sow no.	**Service date**	**Return to service**	**No. born**	**No. weaned**	**Total litter mass weaned**	**REMARKS**

Table 8.5 A sow lifetime record card

Sow no.		Breed								
Date of birth										
Litter	Weaning to service days	Service date	Actual farrowing date	Farrowing interval days	Live births	Still births	No. weaned	Pigs + or −	Litter weaning weight	REMARKS
1										
2										
3										
4										
5										
6										
7										
8										
9										
10										

Table 8.6 A herd performance report form

Data	Month	Month	Month	Six-month average	Suggested target
No. of sows					
No. of gilts					
No. of boars					
No. served to farrow					
No. farrowed					
Farrowing rate					
Pigs born alive					
Pigs born dead					
% born dead					
Total pigs born/sow					
Live pigs born/sow					
Piglet mortality					
Weaner mortality					
Fattener mortality					
Pigs sold					
Pigs born/sow/year					
Pigs sold/sow/year					

Some countries will have figures for average production and targets of production which are relevant to that country. These will be available from government advisory staff or other advisory agencies. This provides the opportunity for the individual producer to assess his pigs' performance against pigs of other producers within his environment.

9 The processing and marketing of pigs

Transport

The stress of transporting pigs to the slaughterhouse can result in pigs dying in transit, dying in lairage at the slaughterhouse, or reduced meat quality in the carcass. The stresses which confront the pig are the handling at loading and unloading, the new surroundings, mixing with strange pigs, the physical discomfort of the journey, and most importantly under tropical conditions, heat stress.

Some measures which can be taken to minimise these stresses are:

- ensure that the loading ramp is properly designed with solid walls and is at the correct height for the cart, truck or trailer.
- handle the pigs quietly and gently at all times. Avoid the use of sticks and prodders.
- do not feed pigs for 12 hours before loading.
- avoid loading and travelling during the heat of the day.
- spray the pigs with cold water before loading and again in the truck.
- provide a cover on the truck, good ventilation, adequate bedding and ensure the floors are not slippery. Make sure that the sides of the truck are high enough to prevent the pigs jumping out. If possible, subdivide animals into groups of 10 or fewer, and never mix pigs of different weight.
- do not stop *en route* to the slaughterhouse.

Lairage

Similar considerations that apply to transporting pigs are also important when the pigs are waiting to be slaughtered. They should be penned under shade in small groups and sprayed with water. They should be fed only if there are long delays before slaughter. Pigs should be handled and driven quietly and gently at all times, and supervised to prevent fighting. As far as possible, and to avoid fear, they should be penned away from the sights and smells of the slaughtering process. All the potential profits which have been achieved during the growing period can be nullified by deaths or damage at this stage.

Slaughter procedures

Stunning

For reasons of animal welfare, pigs should always be stunned before they are bled. Effective stunning ensures prompt and more complete bleeding and also minimises intensive muscle contractions which gives rise to PSE problems (see next section). The main methods of stunning are:

- *Mechanical* A captive-bolt pistol or other implement is used to stun the animal.
- *Electrical* A pair of tongs is used to apply an electrical charge to the pig's head. A current of 1·25 amps and 300 to 600 volts renders the pig unconscious within one second.
- *Gas* Pigs can be led into a tunnel containing 70 to 80 per cent carbon dioxide, when they will lose consciousness within two seconds.

Bleeding

Immediately after stunning the animal should be suspended by its hind legs, and the blood vessels of the neck completely severed to ensure thorough and complete bleeding. The blood should be collected in clean vessels.

Scalding and de-hairing

By immersing the carcass in water at 65 to 75°C, the hair is loosened and can be removed by scraping. Any excess hair can be singed off by a flame.

For the small-scale producer who is slaughtering on the farm, a drum of water over a fire is adequate for scalding purposes. Alternatively, when water is scarce, and if the skins are not used, de-hairing can be achieved by covering the carcass with a 5 cm deep layer of straw or dry grass and burning it. The skin can then be scraped to remove the carbonised surface and any remaining hair.

Evisceration

A long cut is made down the belly from the breast to the hams. To prevent the meat being contaminated, the entire length of the gut should be removed intact. Other internal organs can then be separated, and the gut emptied and cleaned away from the rest of the meat.

Meat hygiene

The freshly-killed carcass is an ideal breeding ground for bacteria, and hygienic conditions are therefore of paramount importance to prevent infections. Ideally, carcasses should be chilled immediately after slaughter, and the meat should remain chilled until it is cooked. Where refrigeration is not available, carcasses should be hung in a cool room, protected from flies by gauze, and then sold and eaten as soon as possible.

At any slaughterhouse, all carcasses should be examined by a qualified meat inspector. He will examine the carcass and offals critically for signs of parasite infection (e.g. 'measly' pork, 'milk-spot' livers, damaged lungs, etc. – see Chapter 7) and other departures from health. Meat that does not pass inspection is condemned and should be burnt.

The carcasses from pigs slaughtered on the farm should also be examined critically so that the transmission of disease and parasites from pigs to humans can be avoided.

Carcass quality

Every country has its own system of assessing the quality of the carcass and generally carcasses are graded according to a number of criteria. These criteria are used to determine the price paid to the producer. The four main considerations are:

- *Conformation* This refers to the shape of the carcass. Carcasses which are poorly developed in the more valuable areas such as the hams and loins are downgraded. Conformation is mainly genetically controlled (see Chapter 2), but can be influenced by nutrition.
- *Degree of fatness* The amount of fat in the carcass is normally estimated by measuring the amount of subcutaneous fat cover at a set position. In Zimbabwe, for example, this is carried out at the 'K' position which is 7·5 cm from the mid-line in line with the last rib (Figure 64). In the UK the 'P2' position is used which is 6·5 cm from the mid-line. The fat measurement which qualifies a carcass for top grade will then be decided according to market demand.
- *Lean content of the carcass* A major fault of carcass grading systems based on measuring the fat cover is that no account is taken of the amount of lean meat in the carcass. In fact, the meatier pigs in a group will often have the heaviest fat cover and will be downgraded. Recently, equipment has been developed which can measure muscle depth in the carcass and therefore be used to estimate the total lean content in the carcass. As a result of these developments, future systems will grade carcasses according to the degree of lean meat in the carcass.

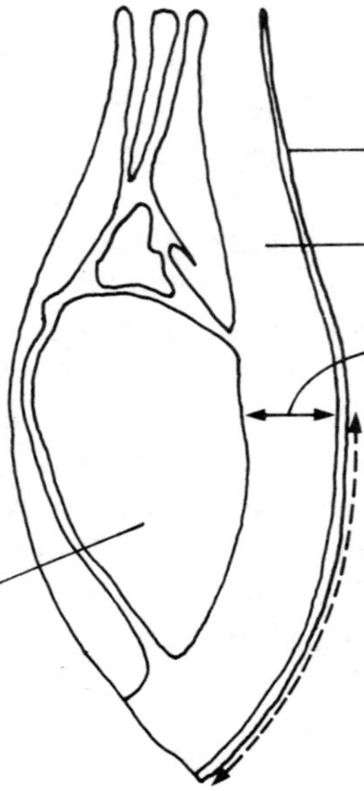

Fig 64 *The 'K' backfat measurement used for grading in Zimbabwe, measured in a cross-section of the half-carcass in line with the tip of the last rib*

- *Fat quality* Carcasses which exhibit a soft and oily, rather than hard, fat cover will tend to be downgraded. This is because they cannot be used for the higher-priced fresh meat or cured products trade. The cause of soft fat can be genetic, but is more likely due to a high proportion of unsaturated fats in the diet. Uneven growth rates, which cause frequent mobilisation and redeposition of fat deposits, can also lead to the condition.

Meat quality

Meat quality, as distinct from carcass quality, relates to the desirability of the meat to the consumer. The most important aspects are:

- *Colour and texture* Pig meat is characteristically paler and more tender than the meat of ruminants, but extremes do occur. Very pale, watery meat is associated with the condition of pale soft exudative (PSE) muscle which occurs in pigs which are homozygous for the 'halothane' gene (see Chapter 4), or as a consequence of poor pre-slaughter handling, or both. Such carcasses show a rapid decrease in acidity at slaughter, which reduces the water-holding capacity of the lean tissue, leading to 'drip-loss' from the meat and excessive paleness. At the other extreme is a condition known as dry, firm and dark (DFD) muscle, which can be attributed directly to pre-slaughter stress. As the age at which pigs are slaughtered increases there is a slow reduction in the paleness and tenderness of the meat produced.

Texture of the meat can be influenced by the amount of intramuscular or 'marbling' fat which is deposited. This can increase from 1 per cent in lean pigs to about 3 per cent in fat pigs, and can improve the succulence and eating quality of the meat. The Duroc breed also tends to have a higher percentage of intramuscular fat than the white breeds. Meat produced from this breed and its crosses will thus have a better meat quality than that from other breeds.

- *Flavour and odour* Pig meat can pick up flavours from the diet, particularly if high levels of fishmeals or rancid fats are fed. However, the main concern is 'boar taint', which can occur in the meat from entire males. The problem does not occur if entire males destined for slaughter are grown out fast enough to allow them to be slaughtered under six months of age.

Marketing

Uses of pig meat

The final phase in pig production is the sale and disposal of the end-product and the pig is extremely versatile in terms of the number of products which can be derived from pig meat. The main categories are:

- *Fresh meat* This is the most important product in the tropics in general, as processing facilities are limited. Pigs destined for the fresh meat trade are usually slaughtered at younger ages and lower weights (porkers) than those used for processing. If sold in the commercial market, the carcass can be butchered into a number of wholesale cuts (see Figure 65) which can be cooked and eaten in a number of different ways.
- *Cured products* These include the various bacons and hams which are cured in brine and can also be flavoured by hanging in smoke. Both processes increase the shelf life of the product compared with fresh meat. Bacon and hams are derived from the 'baconer' category of pig, which is heavier than the porker (see Chapter 8).
- *Other processed products* Included in this category are all the various types of sausages, pies, luncheon meats, hamburgers and meat pastes. These tend to be produced from the lower-value portions of the porker and baconer carcasses, or from heavy hogs, mature sows and boars.

Other pig products in addition to meat are:

- *Lard (pig fat)* This is sold as a separate product for human feed, animal feed or soap production.
- *Pig skin* Pig skin can be made into a valuable leather.

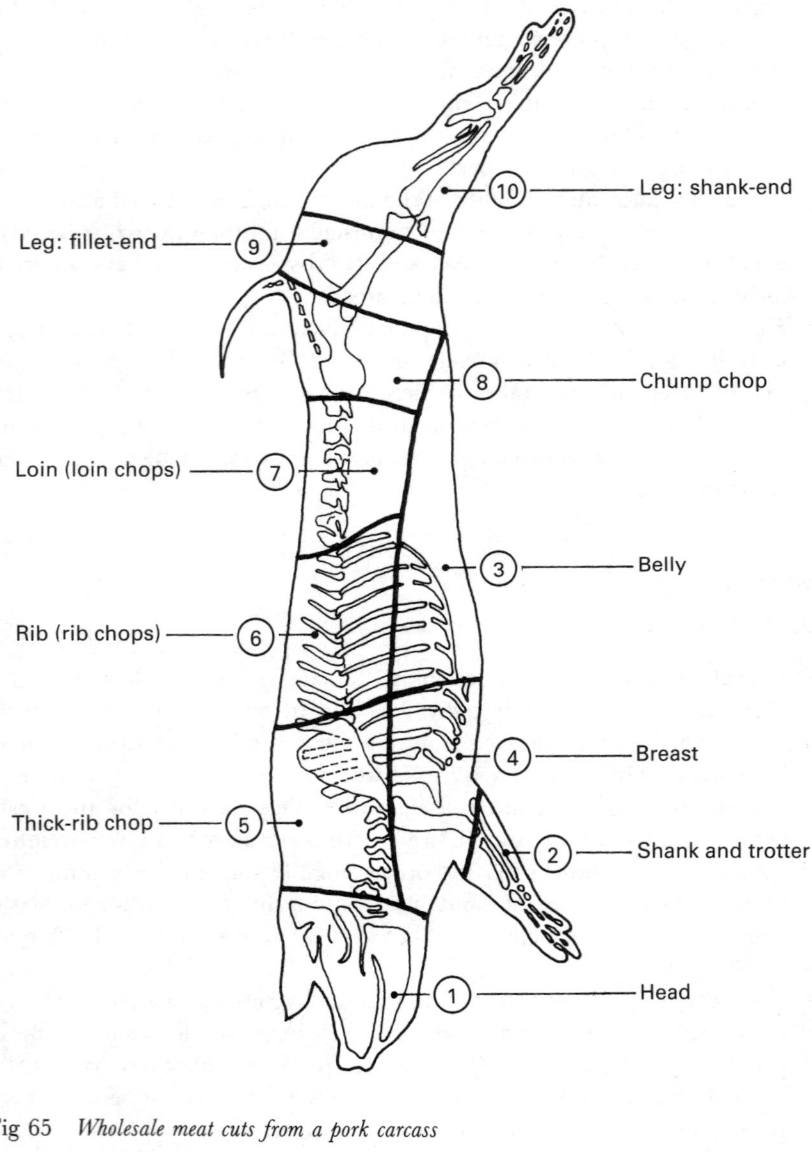

Fig 65 *Wholesale meat cuts from a pork carcass*

- *Bristles* These can be used for brushes and are in especial demand for shaving and paint brushes.
- *Intestines* The intestines are used for sausage casings.
- *Offals* These are all edible and the liver in particular is a delicacy.
- *Blood* Blood is collected separately in many countries and processed into sausages and other delicacies for human consumption.

- *Slaughterhouse by-products* Bones, blood and inedible meat tissue is converted into animal feeds.
- *Hoofs* These are used for gelatin and glue products.

Pig meat as a human food

The specific advantages of pig meat as a human food are often not fully realised by the consumer. The main benefits of pork include the following:

- *Calorific value* Although pork contains a high concentration of nutrients, lean pork is relatively low in calories and of the major meats only chicken has a lower calorific value (see Table 9.1). This allows for a higher intake of the energy-rich, bulkier foods to be enjoyed. These considerations particularly apply to consumers in the developed world, but there is an increasing trend for consumers in the developing world, especially in urban and peri-urban areas, to become more aware of the need to reduce fat in the human diet.
- *Digestibility* Pork is 98 per cent digestible by the human digestive system.
- *Protein quality* Pork contains all the eight essential amino-acids which cannot be manufactured by the human body, and is therefore a high quality, complete protein.
- '*Satiety*' Lean pig meat has been found to have a high 'satiety' value, i.e. it leaves the consumer feeling full for longer periods and helps prevent overeating.
- *Cholestrol* Pork has a relatively low cholestrol level at 0·009% of lean meat.
- *Minerals* Pork has relatively high contents of iron, essential for prevention of anaemia, and zinc, which hastens wound healing, bone growth and tissue development.
- *Vitamins* The levels of the B vitamins are particularly high in pork and it is a leading dietary source of thiamine.

Table 9.1 A comparison of the calorific contents of the main sources of meat (based on an average serving for a meal of 85 g of lean meat)

	No. of calories per 85 gm of lean
Beef	228
Lamb	221
Liver	220
Pork	206
Chicken	144

Note To obtain figure in Kj, multiply by 42.

Systems of marketing

The main systems for marketing pigs are:

- *Private sales* This is the most common method in the tropics among small-scale producers. One, or a number of pigs are sold to local consumers, other producers, butchers or middlemen. The pigs are sold live and the price is generally subject to negotiation. This system has the advantage of being the simplest, but in rural areas individuals who are not aware of current prices can be taken advantage of by speculators and dealers. Due to this problem, marketing co-operatives have been formed in some rural areas to ensure adequate prices for producer members.
- *Public sales* These involve taking the pigs to a central market place, where they are sold by auction on a live basis to the highest bidder. Figure 66 shows a rural pig market in Nigeria.
- *Direct sales to an abattoir or butcher* This method of selling is more applicable to the larger-scale producer. The big disadvantage of direct sales is the effect of the 'pig cycle'. This is the notorious fluctuation in price which occurs in most countries. When pigs are in short supply, prices rise, but this in turn stimulates increased production among producers, and consequently prices fall. As it takes approximately a year for a producer to react to price changes, the cycle will occur every

Fig 66 *A rural pig market in Nigeria*

12 to 18 months. This leads to lack of stability with producers going in and out of pig production.

- *Contract sales* By entering into a contract with an abattoir to supply a certain number of pigs over a period at a set price, the producer is largely protected from the effects of the pig cycle. In turn, this allows him to plan his production output over a longer time.

10 Postscript: Economic and planning considerations

Clearly, the potential profitability of pig production will vary considerably from country to country and between areas within a country. Nevertheless, whether a prospective pig producer in the tropics is planning to be small or large, it is essential that he can answer a few basic questions. Failure to do this will almost invariably lead to loss of money and disillusionment.

The main questions which an aspiring producer needs to ask himself are:

- Have I got an assured market for my pigs and is the price adequate? Can I get my pigs to market without incurring too high a cost, or alternatively, can I site the enterprise closer to the market? Can the market be expanded in the future to provide an increased income?
- Have I sufficient funds to get started? Money is required for fixed capital costs, i.e. buildings and equipment and working capital to pay for stock, feed and labour, for the first year of operation. For a breeding/feeding unit cash outflow tends to increase to a maximum at the end of the first 12 months of operation, while cash inflow starts only 12 months after commencement of the enterprise.
- Can I integrate the pig unit with any other form of production in order to add value to the total enterprise (see Chapter 3)?
- Can I obtain suitable stock for the type of system envisaged?
- Will these supplies be reliable and available at a suitable cost?
- Have I got access to adequate and suitable food resources?
- Are water supplies adequate in quality and quantity?
- Do I enjoy working with animals, particularly pigs? If the answer is yes, this will tend to ensure good stockmanship.
- Have I sufficient management ability? Success in the pig enterprise depends on attention to detail and the labour force in the tropics is often unskilled and untrained.

Only when these questions have been satisfactorily answered should a start be made in pig production.

Appendix A

A guide to the nutrient requirements of pigs (with consideration of the likely availability of food resources in tropical regions)

Concentration in diet (as fed basis)	Growing pigs Body mass range (kg)				Breeding pigs		
	5–10	10–20	20–65	65–120	Pregnant sows	Lactating sows	Boars
Digestible energy (MJ/kg)	13·5	13·0	12·5	12·5	12·5	12·5	12·5
Crude protein (g/kg)	220	190	170	140	120	140	140
Lysine (g/kg)	13	10	9	7	5	7	7
Methionine + cystine (g/kg)	7	5	4·5	4	3·5	4	4
Tryptophan (g/kg)	2	1·9	1·7	1·4	0·8	1·4	1·4
Calcium (g/kg)	7	7	7	5·5	7·5	7·5	5·5
Phosphorus (g/kg)	6	6	5·5	4·5	5	5	4·5
Sodium Chloride (g/kg)	5	5	5	5	5	5	5
Magnesium (g/kg)	0·4	0·4	0·4	0·4	0·4	0·4	0·4
Manganese (mg/kg)	20	20	20	20	20	20	20
Zinc (mg/kg)	100	100	100	100	50	50	100
Iron (mg/kg)	100	100	150	150	80	80	150
Copper (mg/kg)	10	7	7	7	7	7	7
Iodine (mg/kg)	0·14	0·14	0·16	0·16	0·5	0·5	0·2
Selenium (mg/kg)	0·15	0·15	0·15	0·15	0·15	0·15	0·15
Vitamin A (i.u./kg)	6000	6000	4000	4000	6000	4000	4000
Vitamin D_3 (i.u./kg)	2500	2000	1500	1500	2000	1500	1500
Vitamin E (mg/kg)	10	10	10	10	20	20	10
Biotin (mg/kg)	0·3	0·3	0·3	0·3	0·3	0·3	0·3
Riboflavin (mg/kg)	5	4	4	3	4	4	3
Thiamin (mg/kg)	1·3	1·1	1·1	1·1	1·1	1·1	1·1
Vitamin B12 (mg/kg)	18	18	15	15	15	15	15
Pantothenic Acid (mg/kg)	15	15	10	10	15	15	10
Choline (mg/kg)	500	500	500	500	550	550	500

Appendix B

The nutrient composition of the major feedstuffs available for pig-feeding in the tropics (analyses on an 'as–is' basis)

	Macro-nutrients						
	Dry matter %	DE MJ/kg	Crude protein %	Fibre %	Lysine %	Methionine + Cystine %	Tryptophan %
1. Energy sources							
(a) Cereals and by-products							
Maize	89	13·9	8·9	2·9	0·22	0·30	0·09
Maize germ meal	92	11·5	20·0	10·9	0·90	0·85	0·25
Maize bran	89	10·5	10·1	9·0	0·27	0·36	–
Millets	90	12·7	11·5	6·5	0·23	0·48	0·17
Rice	89	10·4	7·3	10·0	0·24	0·22	0·12
Rice bran	89	9·7	13·5	13·0	0·50	0·27	0·10
Rice polishings	90	13·2	11·0	4·0	0·49	0·28	0·14
Sorghum	90	14·2	11·0	2·0	0·27	0·30	0·09
Wheat	89	15·1	10·8	2·8	0·30	0·34	0·12
Wheat bran	89	10·2	14·8	10·0	0·60	0·50	0·30
(b) Other energy sources							
Bananas	22	3·1	1·3	0·8	0·06	0·03	–
Cassava tubers	63	10·5	1·7	5·5	–	–	–
Cassava meal	87	14·6	2·4	7·6	–	–	–
Molasses	74	10·3	2·9	–	–	–	–
Potatoes	23	3·6	2·2	0·70	0·11	0·05	0·02
Sweet Potatoes	32	5·0	1·4	0·53	0·06	0·04	0·03
Yams	25	2·7	2·5	0·70	0·12	0·08	–
2. Protein sources							
Blood meal	89	8·5	80·0	1·0	5·3	2·4	1·0
Chick peas	89	10·8	20·1	9·0	1·4	0·45	0·07
Coconut oil meal	93	11·0	22·0	12·0	0·54	0·53	0·20
Cottonseed meal	91	10·8	41·0	11·0	1·6	1·6	0·70
Feather meal	93	10·0	85·0	1·5	1·1	3·6	0·40
Fishmeal	91	10·2	62·0	1·0	5·3	2·4	0·69
Groundnut meal	92	10·7	48·0	6·8	2·3	1·2	0·39
Kidney beans	89	13·6	25·7	8·2	1·7	0·40	0·70
Linseed oil meal	88	8·8	33·0	9·5	1·1	1·1	0·48

	Macro-nutrients						
	Dry matter %	DE MJ/kg	Crude protein %	Fibre %	Lysine %	Methionine + Cystine %	Tryptophan %
Lucerne meal	92	8·9	20·0	27·0	0·87	0·56	0·46
Meat meal	93	11·2	55·0	2·5	3·0	1·43	0·35
Meat and bone meal	92	10·8	45·0	2·5	2·2	0·95	0·18
Palm kernel meal	90	10·5	17·0	15·0	0·61	0·78	0·17
Pigeon peas	89	11·9	21·0	7·0	1·6	0·31	0·04
Safflower meal	91	12·4	28·5	30·6	0·70	0·68	0·26
Sesame seed meal	94	11·3	42·0	6·5	1·3	1·5	0·82
Skim milk (Dried)	92	14·8	33·0	–	2·6	1·4	0·45
Soyabean meal	89	12·5	46·0	6·0	2·9	1·5	0·55
Sunflower seed meal	93	11·5	42·0	13·0	1·7	1·1	0·50
3. Bulky feeds							
Brewing by-products (dried grains)	92	9·9	27·0	15·0	0·9	1·0	0·34
Banana leaves	20	1·4	2·9	4·5	0·13	0·96	–
Cassava leaves	28	0·83	7·0	3·2	0·43	0·19	0·10
Citrus pulp (dried)	91	8·3	6·0	12·2	0·20	0·19	0·06
Pumpkins & melons	12	1·4	1·5	1·5	–	–	–
Swill (average)	18	3·3	3·6	0·5	–	–	–
Water hyacinth	17	0·31	4·0	2·9	0·14	0·11	0·08

	Minerals								
	Calcium %	Phosphorus %	Sodium chloride %	Magnesium %	Copper mg/kg	Iron mg/kg	Zinc mg/kg	Manganese mg/kg	Iodine mg/kg
1. Energy sources									
(a) Cereals and by-products									
Maize	0·01	0·25	0·08	0·12	5·8	35·0	23·0	5·0	4·5
Maize germ meal	0·04	0·58	0·25	0·30	–	–	18·0	–	–
Maize bran	0·03	0·23	0·05	0·21	–	–	–	0·73	–
Millets	0·05	0·30	0·05	–	–	–	15·0	35·0	–
Rice	0·07	0·31	0·04	0·12	–	20·0	4·0	12·0	4·0
Rice bran	0·10	1·7	0·35	0·95	13·0	190·0	30·0	138·0	–
Rice polishings	0·04	1·4	0·18	0·65	–	160·0	–	–	–
Sorghum	0·04	0·29	0·13	0·13	13·0	53·0	15·0	15·0	0·02
Wheat	0·05	0·30	0·13	0·15	0·04	50·0	15·0	39·0	7·8
Wheat bran	0·14	1·17	0·12	0·55	12·3	200·0	70·0	116·0	12·0
(b) Other energy sources									
Bananas	0·01	0·03	0·19	0·40	–	–	–	–	–
Cassava tubers	0·11	0·06	0·01	0·08	–	–	–	–	–
Cassava meal	0·15	0·08	0·02	0·13	–	–	–	–	–
Molasses	0·82	0·08	–	0·35	59·6	200·0	–	42·2	–
Potatoes	0·01	0·05	0·08	0·03	3·7	–	–	8·8	–
Sweet Potatoes	0·04	0·05	0·04	0·05	–	–	–	–	–
Yams	–	0·04	–	–	–	–	–	–	–

	Minerals								
	Calcium %	Phosphorus %	Sodium chloride %	Magnesium %	Copper mg/kg	Iron mg/kg	Zinc mg/kg	Manganese mg/kg	Iodine mg/kg
2. Protein sources									
Blood meal	0·28	0·22	1·5	0·22	9·9	3000·0	–	5·3	–
Chick peas	0·26	0·32	0·10	–	–	–	30·0	–	–
Coconut oil meal	0·17	0·60	0·10	0·30	–	–	–	–	–
Cottonseed meal	0·18	1·2	0·05	0·54	17·6	100·0	–	23·1	0·12
Feather meal	0·20	0·70	0·71	0·20	–	–	–	21·0	–
Fishmeal	5·0	2·7	1·5	0·20	1·1	270·0	68·0	40·0	20·0
Groundnut meal	0·16	0·52	0·12	0·22	–	20·0	–	18·0	30·0
Kidney beans	0·14	0·54	0·80	0·13	4·1	70·0	42·0	8·4	–
Linseed oil meal	0·35	0·75	–	0·60	25·7	0·03	–	37·6	–
Lucerne meal	1·5	0·27	0·50	0·27	0·55	212·0	35·0	86·0	10·.0
Meat meal	7·6	4·0	3·5	0·27	8·0	410·0	–	18·0	–
Meat and bone meal	11·0	5·9	1·8	1·13	1·3	500·0	98·0	10·0	12·0
Palm kernel meal	0·40	0·50	0·06	0·35	–	–	–	–	–
Pigeon peas	0·24	0·30	0·10	–	–	50·0	30·0	–	–
Safflower meal	0·40	1·1	–	0·37	10·8	560·0	44·0	19·8	–
Sesame seed meal	2·0	1·3	0·10	0·80	–	–	100·0	47·9	–
Skim milk (Dried)	1·3	1·0	0·94	0·11	11·5	50·0	40·0	2·0	–
Soyabean meal	0·25	0·60	0·34	0·28	0·13	150·0	28·5	30·0	70·0
Sunflower seed meal	0·41	0·95	–	–	–	30·0	–	23·0	–
3. Bulky feeds									
Brewing by-products (dried grains)	0·27	0·50	0·26	0·14	21·3	2·5	98·0	37·6	–
Banana leaves	0·11	0·07	–	–	–	–	–	–	–
Cassava leaves	0·14	0·03	0·06	–	–	–	–	–	–
Citrus pulp (dried)	1·4	0·10	0·16	0·16	5·7	1·6	14·5	6·8	–
Pumpkins & melons	–	–	–	–	–	–	–	–	–
Swill (average)	–	–	–	–	–	–	–	–	–
Water hyacinth	–	0·16	–	–	–	–	–	–	–

	Vitamins							
	A i.u./kg	E mg/kg	Biotin mg/kg	Riboflavin mg/kg	Thiamin mg/kg	B12 μg/kg	Panto-thenic acid mg/kg	Choline mg/kg
1. Energy sources								
(a) Cereals and by-products								
Maize	400	22·0	0·23	1·3	4·5	0·21	3·9	400
Maize germ meal	–	91·0	–	3·0	20·0	–	12·0	400
Maize bran	–	–	0·10	1·5	4·4	–	5·3	–
Millets	510	–	–	1·6	6·6	–	7·4	789
Rice	–	–	0·10	1·2	2·4	–	8·6	900
Rice bran	–	1·0	4·2	3·3	5·0	–	14·0	1135
Rice polishings	–	4·0	0·61	11·0	7·3	–	53·0	1237
Sorghum	740	24·0	0·28	1·0	3·6	–	9·0	700
Wheat	250	34·0	0·11	1·0	4·6	1·2	13·0	660
Wheat bran	–	99·0	–	3·1	7·9	–	29·0	1000
(b) Other energy sources								
Bananas	55	–	–	0·50	0·60	–	–	–
Cassava tubers	–	–	–	0·14	–	–	–	–
Cassava meal	–	–	–	0·20	–	–	–	–
Molasses	–	–	–	3·3	0·90	–	38·3	875
Potatoes	–	–	–	0·20	1·5	–	6·4	–
Sweet Potatoes	–	–	–	–	–	–	–	–
Yams	131	–	–	–	–	–	–	–
2. Protein sources								
Blood meal	–	–	0·1	1·5	0·40	42·0	1·2	1000
Chick peas	–	–	0·18	0·40	0·50	–	10·0	642
Coconut oil meal	–	4·0	–	19·8	0·90	–	4·4	–
Cottonseed meal	300	24·0	1·1	5·0	5·3	–	10·3	2300
Feather meal	–	–	0·04	2·1	0·10	0·60	10·0	891
Fishmeal	–	24·0	0·12	6·2	0·90	170·0	8·6	2800
Groundnut meal	300	–	0·40	10·4	0·60	–	48·0	1700
Kidney beans	–	1·0	0·09	1·6	5·5	–	3·0	1670
Linseed oil meal	–	–	–	2·9	9·5	–	–	1225
Lucerne meal	17800	240·0	0·28	13·0	3·0	5·0	26·0	800
Meat meal	–	–	0·09	5·0	0·21	55·0	5·0	2200
Meat and bone meal	–	–	0·10	4·6	0·20	120·0	3·0	1500
Palm kernel meal	69000	–	–	0·5	1·4	–	–	–
Pigeon peas	–	–	0·18	0·8	1·8	–	4·6	642
Safflower meal	550	0·9	1·56	11·3	2·8	–	43·8	2247
Sesame seed meal	–	–	0·34	3·6	2·8	–	6·0	1690
Skim milk (Dried)	24000	9·1	0·33	22·0	3·5	0·01	33·0	1250
Soyabean meal	340	20·0	0·32	26·0	50·0	2·1	14·1	250
Sunflower seed meal	–	11·0	1·45	3·1	–	–	10·0	2894

	Vitamins							
	A i.u./kg	E mg/kg	Biotin mg/kg	Riboflavin mg/kg	Thiamin mg/kg	B12 μg/kg	Pantothenic acid mg/kg	Choline mg/kg
3. Bulky feeds								
Brewing by-products (dried grains)	–	25	0·96	1·5	0·7	–	8·6	1587
Banana leaves	15000	–	–	0·1	0·4	–	–	–
Cassava leaves	–	–	–	0·6	0·4	–	–	–
Citrus pulp (dried)	–	–	–	–	2·4	1·5	13·0	845
Pumpkins & melons	–	–	–	–	–	–	–	–
Swill (average)	–	–	–	–	–	–	–	–
Water hyacinth	–	–	–	–	–	–	–	–

Glossary

Abortion: premature expulsion of the foetus before it is able to survive.
Acaricide: a chemical which kills ticks and mites.
Ad lib (short for *ad libitum*) **feeding:** unrestricted feeding to appetite.
Agalactia: lack of milk or milk let-down in the sow.
Ambient temperature: the temperature in the surrounding environment.
Amino-acids: the basic units of protein.
Anthelminthic: a chemical which kills intestinal worms and parasites.
Antibiotics: a drug, synthesised by micro-organisms, which can inhibit the growth of other micro-organisms.
Antibody: a substance produced in the body in response to invasion by a foreign organism.
Anti-oxidant: a substance that prevents or reduces the oxidation of other substances.
Artificial insemination (AI): the collection of semen from a boar and the introduction of the semen at a later stage into a sow or gilt by means of a catheter.
Bacon: pig meat which has been cured in brine, with or without smoking.
By-product: a product which is additional to the main product(s) for which an animal or process has been developed.
Caecum: part of the hind gut which contains microbes able to digest some fibre in the pig.
Castrate: to remove the testes from a male pig.
Colostrum: the first milk secreted after a sow gives birth.
Concentrate: a feed which contains a high concentration of nutrients, is low in fibre and highly digestible.
Conformation: the physical form and shape of an animal or carcass.
Criss-cross breeding system: when two breeds are crossed, a criss-cross system involves crossing back to each of the parent breeds alternately each subsequent generation.
Diarrhoea: the frequent passage of loose, fluid faeces.
Dressing %: the weight of the carcass expressed as a percentage of liveweight at slaughter.

Endemic disease: a disease regularly found in a specified area or country.
Entire male: a male pig with testes intact (cf. castrate).
Environment: the sum of conditions which affect the well-being of the animal.
Enzyme: a biological catalyst produced by living cells.
Evisceration: removal of the intestines and internal organs.
Farrowing: the act of giving birth to piglets by the sow.
Flushing: increasing the level of feed just before mating in order to stimulate an increase in number of eggs shed.
Food-conversion efficiency: the amount of food required for an animal to make a unit gain in weight.
Gene: a unit of hereditary material in the chromosome.
Genotype: the genetic constitution of an individual.
Gestation: the period from mating to birth of the young. Also known as pregnancy.
Gilt: a young female pig up to the time she produces her first litter.
Growth rate: the liveweight gain of an animal per unit time.
Haemorrhage: profuse bleeding due to rupture of blood vessels.
Heterozygous: in simple Mendelian inheritance, when a character is controlled by a single pair of genes, the heterozygous condition is when both a dominant and recessive gene are present.
Homeotherm (sometimes also called an **endotherm**)**:** a warm-blooded animal.
Homozygous: in simple Mendelian inheritance, when a character is controlled by a single pair of genes, the homozygous condition is when both genes are either dominant (Homozygous dominant) or recessive (Homozygous recessive).
Hybrid: the result of a cross between pure-bred animals.
Hyperthermia: an increase in the deep body temperature of an animal.
Hypothermia: a decrease in the deep body temperature of an animal.
Ill-thrift: failure to thrive normally. Poor condition.
Immunity: the ability to resist and overcome infection.
In utero**:** inside the uterus.
Lacrimation: the secretion of tears from the eyes.
Lactation: the period when the sow is producing milk for her young.
Libido: sexual drive and energy.
Litter: the young produced by a sow at a single birth.
Maintenance level: when the requirements of the animal for nutrients for vital processes when at rest are just met, and there is no gain or loss of nutrients.
Notifiable disease: a disease which must by law be reported to the Government veterinary authorities.
Oedema: accumulation of fluid in the body giving rise to swelling.

Oestrus: the period during which the female will stand and permit the male to mate.

Omnivore: has the ability to eat any type of food.

***Per capita*:** for each person.

Pork: fresh pig meat, normally derived from pigs slaughtered at less than 70 kg liveweight.

Prophylactic: a course of action taken to prevent disease.

Puberty: defines the state when the young animal first becomes capable of breeding.

Rancid: applies to fats and oils which have become oxidised giving a bad smell and taste.

Ruminant: animal possessing a large first stomach known as a rumen, which is capable of digesting large quantities of fibrous material.

Scour: a persistent watery diarrhoea.

Subcutaneous: underneath the skin.

Terminal sire: a boar used solely to produce progeny for slaughter.

Therapeutic: a course of action concerned with treating and curing a disease.

Unthriftiness: a failure to thrive normally – the same as ill-thrift.

Vaccine: a suspension of dead or weakened organisms which stimulate the immune response when introduced into the animal body.

Wallow: a small pond of mud or water in which pigs can roll around to keep cool.

Weaning: the act of separating the young pigs from their mother.

Further reading

AGRICULTURAL RESEARCH COUNCIL (ARC), 1981, *The Nutrient Requirements of Pigs*. CAB: Slough, England.

ANGLADETTE, A., 1966, *Le riz*. Maisonneuve et Larose (Coll. Techniques agricoles et productions tropicales): Paris, pp. 931.

BELLIS, D. B., 1978, *Practical Pig Production in Zimbabwe*, Technical Handbook No. 1. Ministry of Agriculture, Harare, Zimbabwe.

DEVENDRA, C. AND FULLER, M. F., 1979, *Pig Production in the Tropics*. Oxford University Press; Oxford, England.

ENGLISH, P. R., SMITH, W. J. AND MACLEAN, A., 1977, *The Sow–Improving Her Efficiency*. Farming Press: Ipswich, England.

ENGLISH, P. R., FOWLER, V. R., BAXTER, S. AND SMITH, W. J., 1988, *The Growing and Finishing Pig: Improving Efficiency*. Farming Press: Ipswich, England.

EPSTEIN, H., 1971, *The Origin of the Domestic Animals of Africa*, Vol. 2. Africana Publishing Corporation: New York.

EUSEBIO, J. A., 1980, *Pig Production in the Tropics*. Longman: Harlow, England.

FOOD AND AGRICULTURAL ORGANISATION OF THE UNITED NATIONS (FAO), *Quarterly Bulletins of Statistics*, 1988 and 1989. FAO: Rome.

MINISTRY OF AGRICULTURE, FISHERIES AND FOOD, 1982, *Pig Environment*, Booklet No. 2410. MAFF Publications: England.

PATHIRAJA, N., 1986, 'Improvement of Pigmeat Production in Developing Countries. 1. Exploitation of hybrid vigour (heterosis).' *World Animal Review*, *60*, 18–25.

PATHIRAJA, N., 1986, 'Improvement of Pigmeat Production in Developing Countries. 2. Selection Schemes.' *World Animal Review*, *61*, 2–10.

POND, W. G. AND MANER, J. H., 1974, *Swine production in temperate and tropical environments*. W. H. Freeman and Co: San Francisco, USA.

SERRES, H., 1989, Précis d'élevage du porc en zone tropicale. IEMUT. Ministère de la Coopération et du Developpement. La Documentation Française. 2e. éd. pp. 331.

WALTERS, J. R., 1981, 'Peri-urban piggeries in Papua New Guinea', in *Intensive Animal Production in Developing Countries*. BSAP: England.

WHITTEMORE, C. T., 1987, *Elements of Pig Science*. Longman: Harlow, England.

Index

Figures in italics refer to illustrations or tables.